STENNIS SPACE CENTER

TEST FACILITIES CAPABILITY HANDBOOK

(Fourth Edition, Revision 1)

November 2001

Maintained by

New Business Development Office
Propulsion Test Directorate
John C. Stennis Space Center
National Aeronautics and Space Administration
Stennis Space Center, MS 39529-6000

NP-2001-11-00021-SSC

DOCUMENT CHANGE LOG

Status/Change/ Revision	Date of Change	Pg./Para/No. Affected	Description of Change
Initial Release	9/1992	Multiple	Initial printing of the Handbook
First Edition	8/4/94	Multiple	General documentation revision with extensive impact.
Rev. 1	8/22/94	Preface	Preface rewritten
	8/22/94	3.4-1 thru 3.6-10 / 3.4 thru 3.6.7	Section 3.4, "Diagnostic Testbed Facility" deleted. The remainder of Section 3 has been renumbered.
	8/22/94	1-1 thru 1-6/1.0	Introduction rewritten and Sections 1.3.5 and 1.3.6 added.
	8/22/94	Figures	Deleted reference to DTF in all applicable figures.
Rev. 2	11/21/94	5.3.1	Indicated the consolidation of Administrative activities with the IBM 3090 at MSFC.
	11/21/94	5.3.3	Replaced the section titled "Office Automation System."
Second Edition	1/1/95	Multiple	Reformatted & referenced
Rev. 1	8/18/95	Multiple	Revised & updated Sections 3.3 and 3.5
Third Edition	7/1/97	Multiple	Revised to reflect test stand modifications & Technical Support Contract & Facility Operations and Support Services.
Rev. 1	7/30/99	Preface, 1.2, 1.3, 3.3 thru 3.3.3.5.3, 6.2.11, 6.2.12, 6.3.2.4, Appendix A, Appendix B	Revised to reflect E-complex modifications, updated list of contractors, Acronyms and Abbreviations.
Fourth Edition	9/10/01	Multiple	Revised to incorporate current E-complex Facility Capability Documents (FCDs), revised the Propulsion Test Support section, and incorporated general editing changes.
Rev. 1	11/09/01	3.1, 3.3.2.1.1, 3.3.2.1.2.1, 4.0, 4.4, 4.3.1, 5.3.1.1, 5.3.1.1.1, 6.0, 6.1.1, 6.2.2, 6.2.5, 6.2.7, 6.4.5, 5.4.1-5.4.9	Made editorial changes, corrected FOS information, and minor modifications to technical data in the E2, Cell 1 section. Updated TTSC Information Systems section.

PREFACE

Stennis Space Center's (SSC's) test facilities, supporting infrastructure, and technical capabilities are described in this handbook, which should be considered a living and evolving document. Questions related to test capabilities identified in this document and additional assistance in accessibility to the information in this document should be directed to the New Business Development Office as follows:

(228) 688-1646/7244/2001
(228) 688-7885 (FAX)

Or write to:

New Business Development Office
Propulsion Test Directorate
Room 2005B, Building 1100
John C. Stennis Space Center
Stennis Space Center, MS 39529-6000

TABLE OF CONTENTS

STENNIS SPACE CENTER TEST FACILITIES CAPABILITIES HANDBOOK

1.0 INTRODUCTION [9,31]*

The John C. Stennis Space Center (SSC) is located in Southern Mississippi near the Mississippi-Louisiana state line (see Figure 1.1-1). SSC is chartered as the National Aeronautics and Space Administration (NASA) Center of Excellence for large space transportation propulsion system testing. This charter has led to many unique test facilities, capabilities and advanced technologies provided through the supporting infrastructure. SSC has conducted projects in support of such diverse activities as liquid, and hybrid rocket testing and development; material development; non-intrusive plume diagnostics; plume tracking; commercial remote sensing; test technology and more. On May 30, 1996 NASA designated SSC the lead center for rocket propulsion testing, giving the center total responsibility for conducting and/or managing all NASA rocket engine testing. Test services are now available not only for NASA but also for the Department of Defense, other government agencies, academia, and industry.

This handbook was developed to provide a summary of the capabilities that exist within SSC. It is intended as a primary resource document, which will provide the reader with the top-level capabilities and characteristics of the numerous test facilities, test support facilities, laboratories, and services.

Due to the nature of continually evolving programs and test technologies, descriptions of the Center's current capabilities are provided. Periodic updates and revisions of this document will be made to maintain its completeness and accuracy.

1.1 History [9,28]

Selected in October 1961, the Mississippi Test Facility (originally part of the Marshall Space Flight Center), later renamed the National Space Technology Laboratories and known today as the John C. Stennis Space Center, offered ample land for construction of large test facilities with water access for shipping rocket stages and barge loads of propellants. Located in southwest Mississippi near the Mississippi-Louisiana border [approximately 45 miles (mi) from New Orleans, Louisiana], the Center is in a remote, nearly uninhabited area that provides the separation from surrounding communities that is required for test operations. A 13,500-acre fee area contains the facility infrastructure and test facilities for conducting large rocket tests. A 125,327-acre buffer zone, surrounding the fee area, provides an acoustical buffer and a barrier to community encroachment. Refer to Figure 1.1-1 for SSC's regional map.

---

* References are identified in the section headings. Bracketed numbers refer to reference documents and the numbers in parentheses refer to reference drawings, both of which are listed at the end of this handbook.

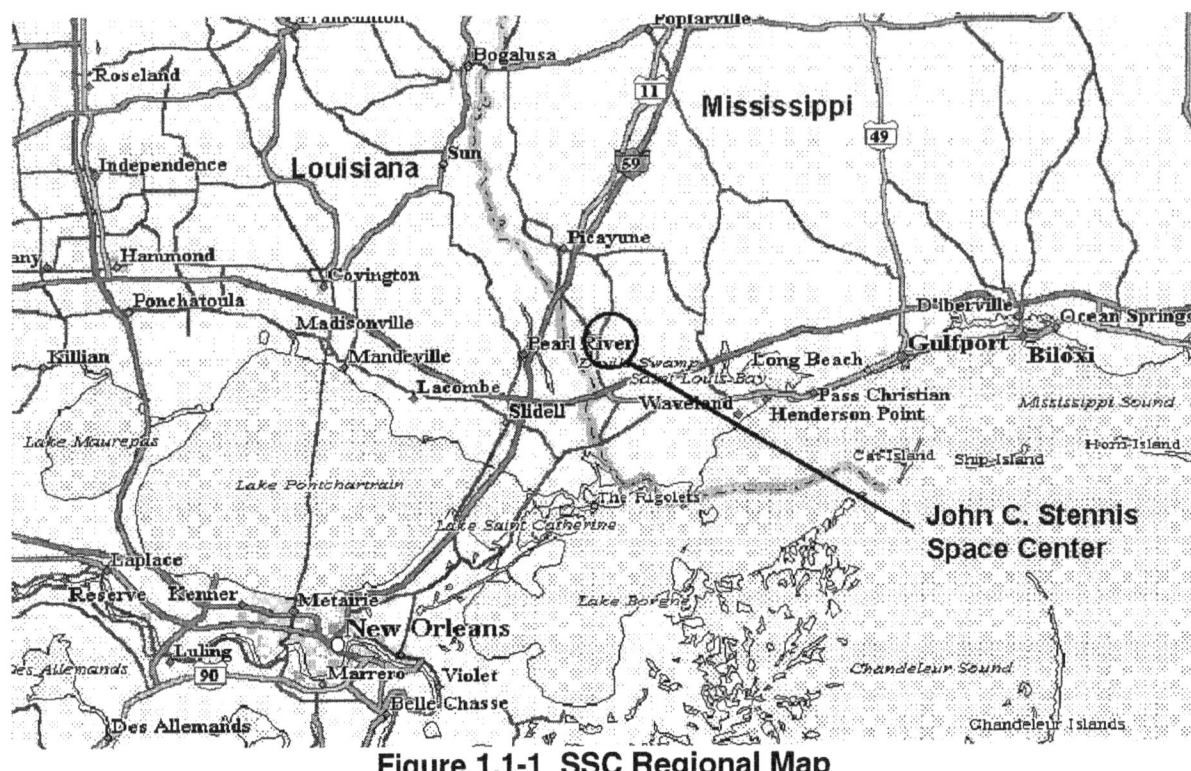

Figure 1.1-1 SSC Regional Map

1.2   Background [9]

Initially established as a national testing center for flight certifying all first and second stages of the Saturn V "Moon Rocket" for the Apollo manned lunar landing program, Stennis Space Center conducted its first static firing on April 23, 1966, and continued testing into the early 1980's. Then, in June 1975, SSC conducted the first test firing of the Space Shuttle Main Engine (SSME). Since then, as the Space Shuttle became an operational, first Generation Reusable Launch Vehicle, SSC has continued testing the SSME. Various improvements and enhancements have been tested which support the evolution of the Space Shuttle. In the late 1980s/early 1990s, several national propulsion development programs, among them the joint DoD/NASA Advanced Launch System (ALS)/National Launch System (NLS) Programs, the Advanced Solid Rocket Motor (ASRM) Program, and the National Aerospace Plane Program, made large investments in ground test infrastructure and capability. One of the results was the construction, activation, and use of the "E" Complex, a major enhancement of the nation's ground test capability. As a result, the E-1, E-2, and E-3 test stands have provided a wide range of test services from small engine systems (~100 lbf thrust) to the large engine systems (650K lbf thrust), from the component level (preburner or gas generator) to the integrated system (rocket engine) level, and from small commercial customers to multi-company, multi-agency consortia.

In addition to its principal mission, SSC has evolved into a multidisciplinary laboratory comprised of other resident federal and state agencies, engaged in space, environmental and national defense programs. NASA, and the other resident agencies share common facilities, services, and capabilities so each may accomplish its own mission in a more efficient and cost-effective manner. SSC has been a leader in the development and improvement of this operationally efficient concept.

1.3 Test Experience

The role of SSC is to provide, maintain, manage, and operate unique test facilities and the related capabilities required for development and acceptance testing and certification of rocket propulsion systems and subsystems. An overview of SSC's test experience is provided in the following sections.

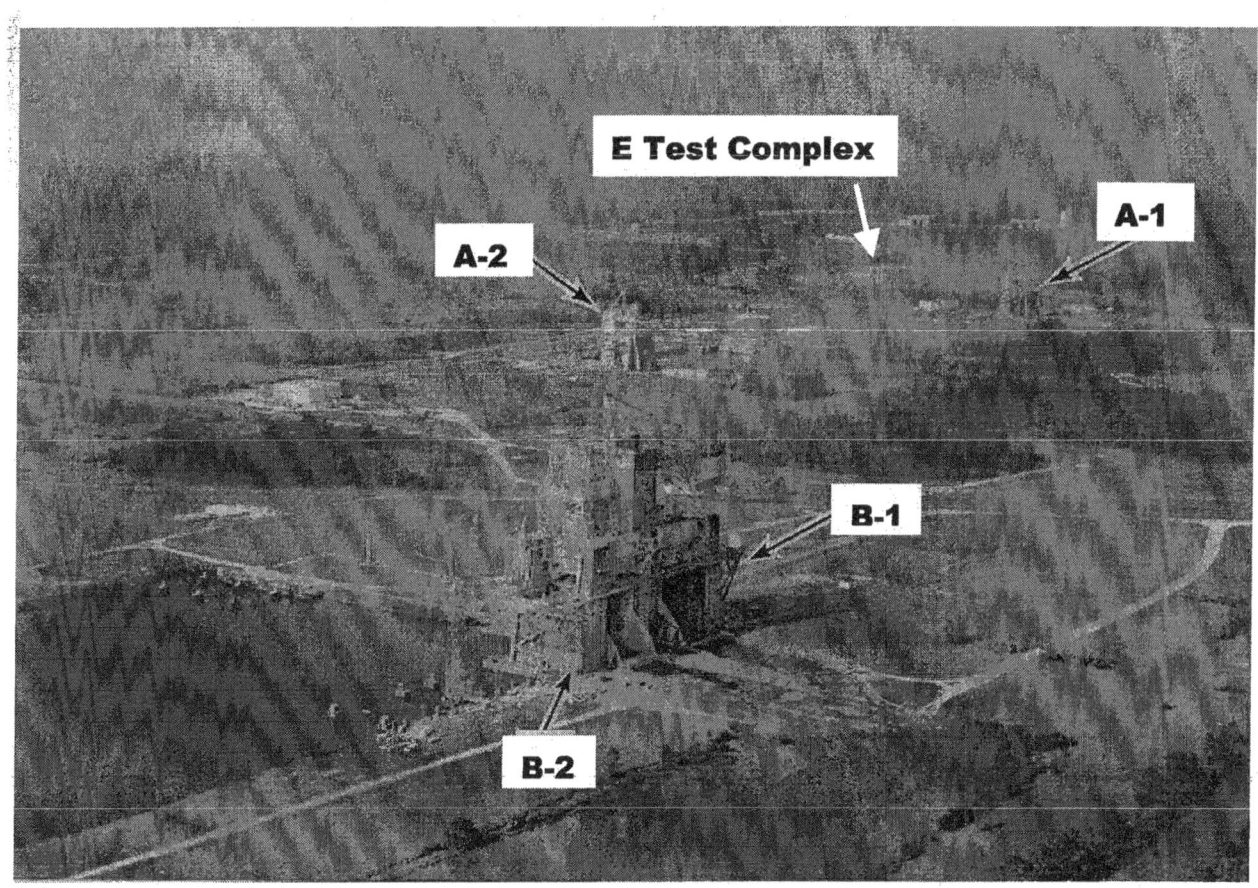

SSC Test Complex

1.3.1 Apollo Saturn V First Stage [33,40]

The Apollo Saturn V First Stage (S-IC) was powered by five F-1 rocket engines capable of a 7.5-million-pound (M-lb) thrust at sea level. The stage was a liquid propellant booster using RP-1 as fuel and liquid oxygen (LOX) as the oxidizer, burning 15 tons of propellant per second. First stages were static fired in the B-2 test position of the "B" Dual Position Test Stand for acceptance testing from March 1967 through September 1970. Fifteen S-IC static firings were conducted for a cumulative duration of approximately 1,875 seconds.

Saturn V First Stage is Lifted Into B-2 Test Stand

1.3.2 Apollo Saturn V Second Stage [23,35,41]

The Apollo Saturn V Second Stage (S-II) consisted of five J-2 rocket engines capable of developing a 1.15-M-lb thrust at altitude. The stage was tested at sea-level conditions, where an 812,500-lb thrust was produced, using liquid hydrogen (LH$_2$) as fuel and LOX as the oxidizer. The first rocket static test firing at SSC was an all-systems test model, the S-II-T, fired in April 1966.

Saturn V Second Stage is Lifted Into A-2 Test Stand

The S-II test program was conducted in the "A" Test Complex's A-1 and A-2 single-position test stands. Fifteen flight stages were static fired for acceptance through October 1970. In all, twenty-seven S-II static firings were conducted for a cumulative duration of approximately 7,106 sec.

1.3.3 Space Shuttle Main Engines

SSME testing began with the first static test firing in May 1975. The SSME flight acceptance testing is currently carried out at SSC to support NASA's Space Transportation System (STS). The SSME develops a 375,000-lb thrust at sea level at a 100% rated power level. At one time, three test stands at SSC were testing SSMEs: A-1 at sea level conditions, and both A-2 and B-1 at simulated altitude conditions.

SSME tests are now conducted on the A-2 test stand. The A-2 test stand can simulate altitude pressure conditions up to 65,000 feet (ft). A total of 2,075 static firings were conducted through 2000 for a cumulative run duration of approximately 697,958 sec. The longest SSME test firings exceeded 2,000 sec. in duration.

1.3.4 Space Shuttle Main Propulsion Test Article (MPTA) [8]

The Space Shuttle MPTA simulated the Space Shuttle main propulsion system. It consisting of three SSME rocket engines, capable of developing a 1.125-M-lb thrust at sea level (100% rated power level). The MPTA static test firings were conducted from April 1978 through January 1981. Eighteen test firings were conducted for a cumulative run duration of approximately 3,775 sec., and were conducted in the B-2 position of the "B" Test Complex dual test stand.

1.3.5 Component Testing

Since initial construction in the late 1980's, various test facilities have provided test services to an increasing number of test programs and test customers.

E-1 Test Facility began its first service testing commercial gaseous and/or liquid oxygen-hybrid rocket motors (1,000 to 10,000 lbs thrust). E-2 Test Facility began service providing research and development testing of prototype hydrogen fuel tanks for the NASP (National Aerospace Plane) Program, as well as testing of prototype "multi-lobe" tanks for the Reusable Launch Vehicle (RLV) Program. The E-2 tank testing took advantage of pressurant and propellant systems in place, as well as added hydraulic loading capability, to test the tanks. In order to provide test service to small test articles with certain pressurants and propellants, the E-3 Test Facility was developed in the 1990's. Hybrid rocket motor testing was conducted in Cell 1, with test article thrust ranging up to approximately 60K lbf. Cell 2 was developed in the late 1990's, providing the nation's most capable ground test facility for testing high concentration hydrogen peroxide/hydrocarbon fueled rocket engine components and systems.

E-1 Test Facility

E-2, Cell 1   **E-3**

1.3.6 Commercial Rocket Engine Component, Engine System, and Booster/Stage Testing

After the ground work was effectively established with the commercial hybrid rocket motor and other commercial test projects in the E-Complex in the mid 1990's, agreements were reached with various commercial companies to expand testing into the A and B Test Complexes. The X-33 prototype vehicle for the RLV Program tested the Aerospike rocket engine on the A-1 test stand, with testing continuing through at least the end of Fiscal Year 2001. Linear Aerospike engine technology was tested, along with other advanced technologies such as Electro-Mechanical Actuators (EMAs). A long-term agreement was reached to allow testing of the commercial RS-68 rocket engine, a large, cryogenic, hydrogen/oxygen rocket engine, on the B-1 test position. A commercially funded modernization provided two test positions on the B-1 test position. The MC-1 rocket engine (formerly known as the FASTRAC engine) was tested on the B-2 test position, in a horizontal mode as well as in a vertical mode on the Propulsion Test Article-1 (PTA-1). Following MC-1 testing, the Common Booster Core (CBC) test article in the Delta IV Program, one of the Evolved Expendable Launch Vehicles (EELV), was tested in the B-2 test position.

1.3.7 Evolution of SSC Into a Multi-Customer Environment

In summary, the number and variety of test projects accomplished at the Stennis Space Center began rapidly expanding in the first half of the 1990's. What historically had been a world-class rocket engine/stage test center dedicated to a single test program, such as the Apollo/Saturn S-IC and S-II stages, and then the Space Shuttle Main Engine/Main Propulsion Test Article, had evolved into a multi-project, multi-program test center. Where the Test Customer had been another NASA Field Center, test customers had become not only other NASA Field Centers, but also other U. S. Government organizations, as well as the Commercial Rocket Propulsion and Space Launch Industry. Then, in 1996, SSC was designated NASA's Lead Center for Rocket Propulsion Testing. By the end of the 1990's and in the year 2000, all of SSC's test stands had active test projects for the first time in the history of SSC.

2.0 NASA TEST OPERATIONS

Test operations are managed, conducted and/or supported, as applicable, in accordance with all applicable project, operations, safety, and quality requirements. The level of management or support provided by NASA is determined through discussions with the New Business Development Office who performs a liaison between potential customers and the functional NASA offices. One of the results of this liaison activity is to clearly delineate the project requirements in a Project Requirements Document (PRD). The PRD is the primary source of requirements by which all other sub-level documents and activities are developed.

2.1 Introduction

Once the testing objectives and requirements are delineated in the PRD, the Propulsion Test Directorate (PTD) is responsible for ensuring the adequate implementation of these test requirements. In this role PTD provides for management of the test facilities and operations required for the testing of propulsion systems, subsystems, components, and related propulsion technologies. To accomplish its management role, PTD performs the functions of Projects Management, System Engineering, Operations, and Test Support.

In addition to PTD, the Safety, Reliability and Quality Assurance (SR&QA) Office is involved in all projects, as applicable, to ensure product assurance and test safety requirements are satisfied. The SR&QA Office provides an independent review capability reporting to the Center Director as well as an in-project support advisory staff as applicable.

2.2 Project Management

Project Management is provided by project offices within PTD, led by a Project Lead/Manager. This function will be resident in either dedicated Project Offices or in the Advanced Projects Office.

2.3 Systems Engineering

The Engineering Division provides the engineering management necessary to ensure that the test facilities and the systems design fulfill the test requirements and safe operations for those test programs assigned to SSC. In this management role, it provides the proper test facility design requirements and incorporation of the test program requirements, including propulsion technology and test studies and data systems design and requirements. Detailed test procedures are reviewed for compliance with program requirements. Support may be provided in the case of non-compliance to ensure that full compliance or acceptable modification of requirements is accomplished. Finally, test requirements are provided to the appropriate test operations organization.

2.4   Test Operations

Carried out by the Operations Division of PTD, Test Operations includes the managing, directing, and safe performance of all test operations accomplished at SSC, by performing component, subsystem, and system level testing. It also ensures facility modifications, trouble-shooting, maintenance, calibration, repair, and retest for test programs. Other operations such as test article and support equipment assembly and handling are also provided as required by a test customer.

2.5   Test Support

The test support function is provided by the A/B/E Complex Division within PTD. This support entails the planning and establishment of requirements for the operation, maintenance, and management of the on-line support facilities and supporting elements. The support elements include providing required pressurants, propellants, industrial water, electrical power, maintenance, work control, and systems engineering support. The expertise for the maintenance, and operation of supporting cryogenics, propellants, pressurants, high-pressure (HP) water, and electrical power systems is also provided as required. Test support is also provided for design of support systems, design reviews, and test support facility configuration baselining.

3.0 PROPULSION TEST FACILITIES

SSC maintains full-scale rocket engine/motor test facilities, component and small engine test facilities and a materials test facility. All of these facilities are described in this section.

The "A" Test Complex and the "B" Test Complex accommodate large test stands sufficient for full-scale, liquid propellant rocket engine and system testing. Located within the E-Complex are three test facilities: E-1, E-2, and E-3. The E-1 facility has 3 test cells that provide ultra high pressure (up to 15,000 psi) testing capability for large-scale propulsion programs. E-2 has similar capability for intermediate sized component and engine test programs. The E-3 facility is capable of performing small and subscale engine/component testing. This facility can also test intermediate size engines in the horizontal position. The capabilities associated with each of these facilities are described in the following sections.

3.1 "A" Test Complex [13,16](85,96-98)

The "A" Test Complex consists of two single-position, vertical-firing test stands designated A-1 and A-2. Each test stand is capable of static firing a test article up to 33 ft in diameter. Each test stand was designed for a maximum dynamic load of 1.1 M-lb vertical (up), 1.7 M-lb vertical (rebound), and 0.7 M-lb (horizontal). Both the A-1 and A-2 test stands are in current/recent operation; thus, the capabilities described at the subsystem level may reflect configurations specific to a particular test program. Test configurations for the test stands have consisted of full flight stage and main propulsion system (ascent vehicle), and single engine testing at sea level and altitude simulation.

The A-1 and A-2 test stands are supplied with cryogenic fluids, hydrogen (GH_2) and inert gases, industrial water, and the electrical power necessary for test operations. LH_2 and LOX are supplied to the stands from cryogenic transportation barges. LH_2 and LOX are supplied to the test article from on-stand run tanks (simultaneous resupply from barge to run tank is possible during extended duration test operations). Gaseous hydrogen (GH_2) is provided as a pressurant for the LH_2 run tank systems. Gaseous nitrogen (GN_2) is provided as a pressurant for the LOX run tank systems.

The A-1 and A-2 test stands are supplied with GH_2 and inert gases from gas storage batteries common to both test stands. The stands are operated from a common Test Control Center (TCC) configured with separate A-1 and A-2 control systems. Both stands also utilize the resources of the Data Acquisition Facility (DAF). Each stand is equipped with a derrick, lifting crane (75 tons, down-rated to 37.5 tons, with a 5-ton jib crane). The cranes are currently downmoded and are not in use. Additional information relating to the "A" Test Complex is contained in the following sections.

11

| A-1 Test Stand | A-2 Test Stand |

3.1.1 Liquid Oxygen Propellant System [37,43](72)

The LOX dock and transfer systems at the A-1 and A-2 test stands provide LOX storage and transfer before, during, and after testing. Each test stand is equipped with a vacuum-jacketed (VJ) LOX run tank, and a downmoded High LOX tank (non-VJ). Relevant data for the existing pressure vessels is presented in the following table.

Vessel Locator Number	Vessel Description	MDWP (psig)	Certified Pressure (psig)	Water Volume (gal)
V-108-LO	A-1 Run Tank	250 (ullage)	250 (ullage)	40,000
V-105-LO	A-2 Run Tank	250 (ullage)	250 (ullage)	40,000
V-103-LO*	A-1 High Tank	125 (ullage)	125 (ullage)	5,500
V-196-LO*	A-2 High Tank	125 (ullage)	125 (ullage)	5,500

*Not in Service

LOX Barge at A-2 Test Stand

LOX propellant subsystem capabilities follow.

      a. Docking facilities for two barges per stand and all shore-to-barge interface accessories, including flexible hoses at the main fill and vent manifolds.

      b. A 10 inch (in.) diameter LOX transfer line from barge dock to test stand, including 40 micron (µ) filtering system; MDWP is 375 pounds per square inch gauge (psig).

      c. A 12 in. diameter LOX supply line from run tank to test article; MDWP is 250 psig.

      d. A 12 in. diameter LOX dump line rated at 4220 gallons per minute (gal/min). This line transfers LOX to the LOX dump pit during chilldown bleeds and emergency situations.

      e. A dockside, deluge-water piping system supplying industrial water at 5,000 gal/min and 215 psig.

3.1.2 Liquid Hydrogen Propellant System [37,43]
(12, 13, 17, 61, 62, 73, 82-84, 100)

The LH_2 dock and transfer systems at the A-1 and A-2 test stands provide LH_2 storage and transfer before, during, and after testing. Each test stand is equipped with a VJ LH_2 run tank. Relevant data for the existing pressure vessels is presented in the following table.

Vessel Locator Number	Vessel Description	MDWP (psig)	Certified Water Pressure (psig)	Volume (gal)
V-107-LH	A-1 Run Tank	50 (ullage)	50 (ullage)	110,000
V-104-LH	A-2 Run Tank	50 (ullage)	50 (ullage)	110,000

LH_2 propellant subsystem capabilities follow.

a. Docking facilities for two LH_2 barges per test stand and all shore-to-barge interface accessories, including flexible hoses at the main fill and vent manifolds.

b. A 12 in. diameter, VJ LH_2 transfer line from barge dock to test stand rated at 70 psig, including a 200 µ filtering system.

c. A 12 in. diameter, VJ LH_2 supply line from run tank to test article, with maximum operating pressure of 65 psig.

d. A 12 in. diameter, LH_2 barge, H_2 vent line to flare stack for LH_2 barge boil-off.

e. A 24 in. diameter, test stand, H_2 vent line to flare stack for run tank LH_2 boil-off and line purge venting.

f. A dockside, deluge-water piping system supplying industrial water at 8,000 gal/min, and 215 psig.

3.1.3 Gas Supply Systems [37,43]

The high-pressure, inert-gas battery in the "A" Test Complex contains air, helium (He), and nitrogen (N_2) gases. A separate gas battery is utilized for GH_2 storage. Gas batteries are common to both test stands, and they function as accumulators that are resupplied by cross-country gas transfer from the High Pressure Gas Facility (HPGF).

3.1.3.1 Air Supply System (47-50,70,71,80,81)

The HP air storage, in the "A" Test Complex gas storage battery, consists of three HP vessels containing missile-grade air. Relevant data for the existing pressure vessels is presented in the table on the next page.

Vessel Locator Number	MDWP (psig)	Certified Pressure (psig)	Water Volume (ft³)	Storage at 2800 psig, 70 °F (scf)
V-096-HA	3,750	2,879	950	176,802
V-098-HA	6,300	4,647	1,500	279,161
V-099-HA	6,300	4,556	1,500	279,161

Design and operating characteristics of the HP air transfer systems are:

a. A 3 in. diameter transfer line from HPGF to Valve Pit 5 (VP-5); with operating pressure of 2,800 psig and allowable pressure drop of 200 psig, MDWP is 3,700 psig

b. A 2 in. diameter transfer line from VP-5 to VP-15; with an operating pressure of 2,800 psig and allowable pressure drop of 200 psig, MDWP is 3,700 psig.

c. A 1 ½ in. diameter transfer line from VP-15 to "A" Test Complex gas storage battery; with operating pressure of 2,800 psig and allowable pressure drop of 200 psig, MDWP is 3,700 psig

d. A 2 in. diameter transfer line from "A" Test Complex gas storage battery to A-1 and A-2 test stands; with operating pressure of 2,800 psig and allowable pressure drop of 200 psig, MDWP is 3,500 psig.

3.1.3.2 Helium Supply System (42-45,66,67,76,77)

The HP gaseous helium (GHe) storage at the "A" Test Complex gas storage battery consists of one HP vessel. Relevant data for the existing pressure vessel is presented in the following table.

Vessel Locator Number	MDWP (psig)	Certified Pressure (psig)	Water Volume (ft³)	Storage at 3000 psig, 70 °F (scf)
V-101-HE	6,300	4,753	1,500	279,996

Design and operating characteristics of the HP GHe transfer systems are:

   a. A 1 ½ in. diameter transfer line from HPGF to VP-7; with operating pressure of 3,000 psig, MDWP is 6,300 psig

   b. A 1 ½ in. diameter transfer line from VP-7 to A-2 Pressure Reducing Area (PRA); with operating pressure of 3,000 psig, MDWP is 6,300 psig

   c. A 4 in. diameter transfer line from A-2 PRA to GHe gas battery and A-1 PRA; with operating pressure of 3,000 psig, MDWP is 6,000 psig

   d. A 3 in. diameter transfer line from A-2 PRA to A-2 test stand; with operating pressure of 3,000 psig and allowable pressure drop of 400 psig, MDWP is 3,700 psig

   e. A 3 in. diameter transfer line from A-1 PRA to A-1 test stand; with operating pressure of 3,000 psig and allowable pressure drop of 400 psig, MDWP is 3,700 psig.

3.1.3.3   Nitrogen Supply System (32-35,39,63-65,74,75)

The HP GN_2 storage in the "A" Test Complex gas storage battery consists of four HP vessels. Relevant data for the existing pressure vessels is presented in the following table.

Vessel Locator Number	MDWP (psig)	Certified Pressure (ft³)	Water Volume	Storage at 4400 psig, 70 °F (scf)
V-068-GN	6,300	4,500	1,500	393,642
V-082-GN	6,300	4,576	1,500	393,642
V-097-GN	6,300	4,578	1,500	393,642
V-100-GN	6,300	4,678	1,500	393,642

Design and operating characteristics of the HP GN_2 transfer systems are:

   a. A 4 in. diameter transfer line from HPGF to A-2 PRA; with operating pressure of 4,400 psig and allowable pressure drop of 100 psig, MDWP is 6,300 psig

   b. An 8 in. diameter transfer line from A-2 PRA to GN_2 gas battery and A-1 PRA; with operating pressure of 4,400 psig and allowable pressure drop of 100 psig, MDWP is 6,300 psig

   c. A 6 in. diameter transfer line from A-2 PRA to A-2 test stand; with operating pressure of 3,000 psig and allowable pressure drop of 100 psig, MDWP is 3,000 psig.

d. A 6 in. diameter transfer line from A-1 PRA to A-1 test stand; with operating pressure of 3,000 psig and allowable pressure drop of 100 psig, MDWP is 3,000 psig.

3.1.3.4   Hydrogen Supply System (10-12,14-16,27,68,69,78,79)

The HP GH$_2$ storage area at the "A" Test Complex consists of four HP vessels. Relevant data for the existing pressure vessels is presented in the following table.

Vessel Locator Number	MDWP (psig)	Certified Pressure (psig)	Water Volume (ft^3)	Storage at 3000 psig, 70 °F (scf)
V-066-GH	5,000	—*	1,250	226,963
V-071-GH	5,000	—*	1,250	226,963
V-219-GH	5,000	5,000	600	108,942
V-223-GH	5,000	5,000	600	108,942

*Not in Service

Design and operating characteristics of the HP GH$_2$ transfer systems are:

a. A 2 in. diameter transfer line from HPGF to A-2 PRA; with operating pressure of 3,000 psig and allowable pressure drop of 200 psig, MDWP is 6,000 psig

b. A 2 in. diameter transfer line from A-2 PRA to GH$_2$ gas battery and A-1 PRA; with operating pressure of 3,000 psig and allowable pressure drop of 175 psig, MDWP is 6,000 psig

c. A 1 ½ in. diameter transfer line from A-2 PRA to A-2 test stand; with operating pressure of 3,000 psig and filtered to 10 µ, MDWP is 3,700 psig

d. A 1 ½ in. diameter transfer line from A-1 PRA to A-1 test stand; with operating pressure of 3,000 psig and filtered to 10 µ, MDWP is 3,700 psig

3.1.4 Water Supply System [7]

Water for the "A" Test Complex is supplied from the High Pressure Industrial Water (HPIW) facility at a pressure of 215 psig, through a 75 in. diameter water line, at an operating flow rate of 212,000 gal/min. At the "A" Test Complex, the 75 in. diameter line is branched to two 66 in. diameter lines supplying the A-1 and A-2 test stands. See section 4.3.1 for additional information related to HPIW support operations.

High Pressure Water Facility

3.1.5 Electrical Power Distribution [7](51)

Electrical power is distributed to the entire Test Complex by two, 13.8 kilovolt (kV) feeder circuits from the SSC Main Substation. At the "A" Test Complex, circuits 11 and 21 source double-ended, 480 volt (V) substations at the A-1 and A-2 test stands, the DAF, and the "A" Test Control Center (A-TCC), all of which step 13.8 kV down to 480 V. Feeder circuit breakers then distribute the 480 V to lighting panels, Motor Control Centers (MCCs), transformers, 28 volt direct current (Vdc) rectifiers, etc., for utilization. Control power is provided by redundant 28 Vdc rectifiers with battery backup should the 480 V fail. All safety-critical control devices are powered with 28 Vdc to ensure that a test can be terminated safely if there is a power failure.

During engine testing, the test conductor may request that the Emergency Generators (refer to Section 4.3.2) are running and ready to supply power to the Test Complex should utility power fail.

3.1.6 Data Acquisition [32]

The Data Acquisition Facility (DAF) is the central data collection and processing facility receiving engine pre- and post-test data from the test article and the A-Test Control Center (A-TCC). The DAF uses a host computer to acquire, process, store, and display real-time and post-test data. The low-speed Data Acquisition System (DAS) acquires 400 analog input channels per test stand (1,200 channel total) at a rate of 50 samples per second (sps) each. Another data acquisition computer acquires up to 180 analog input channels at a maximum sample rate of 20,000 sps, with a throughput of about

18

450,000 sps for a 1000 sec. test. Patch panels are used to connect the current test stand (A-1, A-2, or B-1) to this DAS. High-Frequency (HF) analog tape recorders record up to a 180 signal total for post-test digitizing and analysis. Front-end signal conditioning and amplifiers for Resistance Temperature Detectors (RTDs), thermocouples, strain gages, etc. are provided as required at each test stand. The operator graphics display system displays real-time measurements and can provide red-line cutoffs for critical measurements. A modernization project upgraded and increased the capacity of the DAS in 1997. This project provided greater flexibility and independence to each test position to significantly enhanced data processing and test turnaround.

3.1.7 Test Control [32]

The A-TCC is a dual control room facility (for the A-1 and A-2 test stands), which houses Control System (CS) equipment and serves as the central command location for both test conductor and test personnel, who are required to observe, monitor, supervise, and control test operations at the A-1 or A-2 test stand during a test firing. Each TCC area contains a computer for data collection; a facility control console with meters, indicator lights and operating devices; Closed Circuit Television (CCTV) monitors for test stand surveillance; audio communications; and a graphics display instrumentation system. The manually operated control system monitors critical test functions and can initiate corrective actions or terminate a test.

3.2 "B" Test Complex [15,34](95)(99)

The "B" Test Complex consists of a dual-position, vertical, static-firing test stand designated the B-1/B-2 test stand. Each test position is capable of static-firing test articles up to 33 ft. in diameter. Each position is designed for a maximum dynamic load of 11.0 M-lb vertical (up), 8.5 M-lb vertical (rebound), and 6.0 M-lb (horizontal). The B-1 test position is in operation for RS-68 testing, thus, the capabilities described at the subsystem level may reflect configurations specific to that test program. The B-2 test position completed the Delta IV Common Booster Core Program in fiscal year 2001, and thus, the capabilities described at the subsystem level may reflect configurations specific to that test program.

The B-1/B-2 test stand is capable of being supplied with LOX and LH_2 cryogenic fluids, GH_2, and inert gases; industrial water; and the electrical power necessary for test operations. LH_2 and LOX are supplied to the stands from cryogenic transportation barges and supplied to the test article from on-stand run tanks. (Resupply from barge to run tank is possible during test operations for extended duration testing.) GH_2 is provided as a pressurant for the LH_2 barge and the run tank systems. GN_2 is provided as a pressurant for the LOX run tank systems.

B-1/B-2 Test Stand

The B-1/B-2 test stand is supplied with GH$_2$ and inert gases from "B" Test Complex gas storage batteries. The test stand's two positions are operated from a common TCC configured with separate B-1 and B-2 control systems. Both positions have separate Dedicated Acquisition and Control Systems (DACS). They share common DAF resources, which are also utilized by the A-1 and A-2 test stands. The B-1/B-2 test stand is equipped with a 200-ton, main derrick, lifting crane (currently proof tested to 37.5 tons), with a 20-ton jib crane and a 175-ton auxiliary derrick, lifting crane (currently proof tested to 37.5 tons). Additional information relating to the "B" Test Complex is contained in the sections that follow.

3.2.1 Liquid Oxygen Propellant System [37,43](86,93)

Before, during, and after testing, the LOX dock and transfer systems at the "B" Test Complex provide LOX storage and transfer for both the B-1 and B-2 test positions. The B-1 test position is equipped with a VJ LOX run tank. B-2 is also equipped with a VJ LOX storage/transfer tank. Relevant data for the existing pressure vessels is presented in the following table.

Vessel Locator Number	Vessel Description	MDWP (psig)	Certified Pressure (psig)	Water Volume (gal)
V-150-LO	B-1 Run Tank	110 (ullage)	110* (ullage)	49,550
V-266-LO	B-2 Storage	146 (ullage)	135 (ullage)	28,000

Maximum operating pressure of 140 psig (ullage).

3.2.1.1 B-1 LOX Propellant Subsystem

   a. Dock facilities for three (3) LOX barges and all shore-to-barge interface accessories, including flexible hoses at main fill-and-topping manifolds. Two additional barge positions are available but have no piping or mooring devices.

   b. A 14 in. diameter LOX transfer line (from barge dock to dual-stand B-1 position) with operating pressure of 310 psig. The line is reduced to a 10 in. diameter at the test stand and has 40 μ filtering system supply.

   c. A 12 in. diameter LOX supply line from run tank to test article, with MDWP of 250 psig.

   d. A 4 in. diameter topping line between two (2) dock positions (No. 1 & No. 2) and dual test stand, with operating pressure of 275 psig. The line is reduced to 3 in. diameter at the test stand, which supplies the B-2 test position.

   e. An 18 in. diameter LOX dump line to deliver LOX to LOX dump pit during chilldown bleeds and emergency situations.

f. A dockside, deluge-water piping system supplying industrial water at 70,000 gal/min, 215 psig.

3.2.1.2   B-2 LOX Propellant Subsystem

a. B-2 test stand has a 28,000 gallon storage tank for transfer of LOX to the test article run tank.

b. 4 in. piping to Level 7 and 3 in. from Level 7 to Level 8.

c. Capable of transferring LOX at a flow rate of 0. lb/s to 20 lb/s.

3.2.2   Liquid Hydrogen Propellant System [37,43]

3.2.2.1   B Complex Propellant Subsystem

Before, during, and after testing, the LH_2 dock and transfer systems at the "B" Test Complex provide LH_2 storage and transfer for both the B-1 and B-2 test positions. The B-1 test position is equipped with a VJ LH_2 run tank. Relevant data for the existing pressure vessel is presented in the following table.

Vessel Locator Number	Vessel Description	MDWP (psig)	Certified Pressure (psig)	Water Volume (gal)
V-151-LHB-1	Run Tank	66 (ullage)	66 (ullage)	90,000

LH_2 propellant subsystem capabilities are as follows:

a. Dock facilities for three LH_2 barges and all shore-to-barge interface accessories, including flexible hoses at main fill and LH_2 barge H_2 vent manifolds.

b. A 10 in. diameter, VJ, LH_2 transfer line from barge dock to dual-stand B-1/B-2 positions, with operating pressure of 70 psig and including 200 μ filtering system.

c. A 12 in. diameter, VJ, LH_2 supply line from run tank to test article, with MDWP of 100 psig.

d. A 12 in. diameter, LH_2 barge, H_2 vent line to flare stack for LH_2 boil-off.

e. An 18 in. diameter, test stand, H_2 vent line to flare stack for run tank boil-off and line purge venting.

f. A dockside, deluge-water piping system supplying industrial water at 12,500 gal/min, 215 psig.

3.2.3 Gas Supply Systems [37,43]

The HP inert-gas battery in the "B" Test Complex contains HPA, He, and GN_2 gases. A separate gas battery is utilized for GH_2. The gas batteries serve the B-1 and B-2 test positions and function as accumulators that are resupplied by cross-country gas transfer from the HPGF.

3.2.3.1   Air Supply System (91, 94)

The HP air storage at the "B" Test Complex inert gas storage battery consists of three HP vessels containing missile-grade air. Relevant data for the existing pressure vessels is presented in the following table.

Vessel Locator Number	MDWP (psig)	Certified Pressure (psig)	Water Volume (ft³)	Storage at 4400 psig, 70 °F (scf)
V-141-HA	6,300	4,814	1,065	198,204
V-143-HA	6,300	4,693	1,500	279,161
V-146-HA	3,750	2,917	950	176,802

Design and operating characteristics of the HP Air transfer systems are:

a. A 3 in. diameter transfer line from HPGF to VP-5; with operating pressure of 2,800 psig and allowable pressure drop of 200 psig, MDWP is 3,700 psig.

b. A 2 in. diameter transfer line from VP-5 to VP-6; with operating pressure of 2,800 psig and allowable pressure drop of 200 psig, MDWP is 3,700 psig.

c. A 1 ½ in. diameter transfer line from VP-6 to "B" Complex gas storage battery; with operating pressure of 2,800 psig and allowable pressure drop of 200 psig, MDWP is 3,700 psig.

d. A 2 in. diameter transfer line from gas battery to B-1 test position; with operating pressure of 2,800 psig and allowable pressure drop of 400 psig, MDWP is 3,700psig.

3.2.3.2 Helium Supply System (92)

The HP GHe storage at the "B" Test Complex inert gas storage battery consists of one HP vessel. Relevant data for the existing pressure vessel is presented in the following table.

Vessel Locator Number	MDWP (psig)	Certified Pressure (psig)	Water Volume (ft³)	Storage at 3000 psig, 70 °F (scf)
V-144-HE	6,300	4,725	1,500	279,996

Design and operating characteristics of the HP GHe transfer systems are:

a. A 1 ½ in. diameter transfer line from HPGF to VP-7; with operating pressure of 3,000 psig, MDWP is 6,300 psig.

b. A 1 ½ in. diameter transfer line from VP-7 to "B" Complex gas storage battery; with operating pressure of 3,000 psig and allowable pressure drop of 200 psig, MDWP is 6,300 psig.

c. A 3 in. diameter transfer line from gas battery to B-1 test position; with operating pressure of 3,000 psig and allowable pressure drop of 300 psig, MDWP is 6,300 psig.

3.2.3.3 Nitrogen Supply System (89-90)

The HP GN_2 storage at the "B" Test Complex inert gas storage battery consists of four HP vessels. Relevant data for the existing pressure vessels is presented in the following table.

Vessel Locator Number	MDWP (psig)	Certified Pressure (psig)	Water Volume (ft³)	Storage at 4400 psig, 70 °F (scf)
V-225-GN6600	6,600	6,600	750	196,821
V-226-GN6600	6,600	6,600	750	196,821
V-231-GN6600	6,600	6,600	750	196,821
V-232-GN6600	6,600	6,600	750	196,821

Design and operating characteristics of the HP GN_2 transfer systems are:

a. A 4 in. diameter transfer line from HPGF to VP-7; with operating pressure of 4,400 psig and allowable pressure drop of 100 psig, MDWP is 6,300 psig.

b. A 3 in. diameter transfer line from VP-7 to "B" Test Complex GN_2 storage battery; with operating pressure of 4,400 psig, MDWP is 6,300 psig.

c. An 8 in. diameter transfer line from "B" Test Complex GN_2 storage battery to test stand; with operating pressure of 4,400 psig and allowable pressure drop of 100 psig, MDWP is 6,300 psig.

3.2.3.4 Hydrogen Supply System

The HP GH_2 storage area at the "B" Test Complex consists of three HP vessels. Relevant data for the existing pressure vessels is presented in the following table.

Vessel Locator Number	MDWP (psig)	Certified Pressure (psig)	Water Volume (ft³)	Storage at 3000 psig, 70 °F (scf)
V-224-GH	5,000	5,000	600	108,942
V-227-GH	6,600	6,600	608.9	110,558
V-072-GH (Rail-Car)	5,000	3,375	2,500	453,925

Design and operating characteristics of the HP GH_2 transfer systems are:

a. A 2 in. diameter transfer line from HPGF to Test Complex (tee joint to "A" and "B" Test Complexes); with operating pressure of 3,000 psig and allowable pressure drop of 200 psig, MDWP is 6,000 psig.

b. 1 ½ in. diameter transfer line from Test Complex (tee joint "B" Test Complexes) to "B" Test Complex GH_2 storage battery; with operating pressure of 3,000 psig, MDWP is 3,200 psig.

c. A 2 in. diameter transfer line from "B" Test Complex, GH_2 storage battery to test stand; with operating pressure of 3,000 psig and allowable pressure drop of 175 psig, MDWP is 4,400 psig.

3.2.4 Water Supply System [7]

Water for the "B" Test Complex is supplied from the HPIW facility at a pressure of 215 psig, through a 96 in. diameter water line at an operating flow rate of 330,000 gal/min. See section 4.3.1 for the HPIW support operations.

3.2.5 Electrical Power Distribution [7](51)

Electrical power is distributed to the Test Complex by two, 13.8-kV feeder circuits from the SSC Main Substation. At the "B" Test Complex, circuits 11 and 21 source double-ended, 480 V substations at the B-1/B-2 dual test stand and at the "B" Complex TCC, and step the 13.8 kV down to 480 V. Feeder circuit breakers then distribute the 480 V to lighting panels, MCCs, transformers, 28-Vdc rectifiers, etc., for utilization. Control power is provided by redundant 28 Vdc rectifiers with battery backup should the 480 V fail. All safety-critical control devices (PLCs, relays, solenoid valves, etc.) are powered with 28 Vdc to ensure that a test can be terminated safely if there is a power failure. During engine testing, the test conductor may request that the Emergency Generators (see Section 4.3.2) are running and ready to supply power to the Test Complex should utility power fail.

3.2.6 Data Acquisition [32]

The B-TCC is the central data collection and processing facility receiving engine pre- and post-test data from the test article and the B test facility. The B-TCC uses redundant loggers connected to an isolated network to acquire, store, and post-process test data. Realtime display of data is provided in both tabular form as well as graphical trend. The low-speed DAS acquires up to 592 analog input channels at a rate of 250 sps each. Another data acquisition computer acquires up to 120 analog input channels at a maximum sample rate of 100,000 sps. Patch panels are used to connect the current test stand to this low-speed DAS. Front-end signal conditioning and amplifiers for RTDs, thermocouples, strain gages, etc., are provided, as required, at the test stand. Redundant signal outputs are patched to an offline computer system to provide red-line cutoffs for critical measurements.

A modernization project upgraded and increased the capacity of the DACS in 1997. Test stands A-1, A-2, and B-1 have separate, dedicated systems with the capabilities listed in Table 3.2.6-1. These systems have a very low setup time, reconfiguration, and pretest/post test tasks.

1.	512 Channels of Low Speed Analog Data with signal conditioning and A/D Mux=s
2.	128 Channels of Digital High Speed Analog Data recorded on 8 MM tape cassettes
3.	High Speed Fiber optic data network between test stands and Test Control Centers (TCC=s)
4.	Automated system configuration, setup, and checkout procedures
5.	Computer based Control System with Graphical User Interface (GUI) for facility functions

Table 3.2.6-1 Test Complex Data Acquisition and Control System Capabilities

The B-2 DAS has signal conditioning, amplification, and multiplexing for 256 active channels with a 320 channel capacity on the 10th floor of the test stand. These consist of low frequency (100 samples per second) and high frequency (100,000 samples per second) channels with up to 128 channels for high frequency signals. The B-2 DAS also provides digital interfaces for the acquisition of discrete events. These channels are digitized and transmitted via fiber optics to the B-2 control room in the B-TCC where they are recorded and displayed. An interface is provided to the Control System for monitoring selected DAS signals for blueline/redline limit checking.

3.2.7 Test Control [32]

The B-TCC is a dual control room facility for the B-1 and B-2 test positions. This facility houses CS equipment and serves as the central command location for the test conductor and personnel, who are required to observe, monitor, supervise, and control test operations at the B-1 (or B-2) test position during a test firing. Each TCC area contains a computer for data collection; a facility control console with meters, indicator lights, and operating devices; high/low speed CCTV cameras, recorders, and monitors for test stand surveillance; audio communications; and a graphics display instrumentation system. Separate B-1/B-2 operated control systems monitor critical test functions and initiate corrective actions or terminate a test.

3.3 "E" Test Complex [3](56-59)

The E Test Complex consists of a high flow rate facility, high heat flux facility and a subscale test facility. The E-1 multi-cell test facility is a high flow rate facility capable of testing gas generator and preburner driven liquid oxygen/liquid hydrogen (LOX/LH$_2$) turbopump assemblies. It is also capable of testing a variety of combustion devices and other rocket engine components and sub-assemblies. The E-2 multi-cell test facility is a versatile test complex for developmental testing projects involving hot gas, cryogenic fluids, gas impingement, inert gases, industrial gas, specialized gases, hydraulics, and deionized and industrial water. The E-3 Test Facility is a versatile test complex that is available for component development testing of combustion devices, rocket engine components and small/subscale complete engines and boosters. Particularly unique is the established capability at E-3 to test components and rocket engines using high concentration (up to 98%) hydrogen peroxide. The facility has two test cells, Cell 1 is a horizontal test stand, which is capable of testing intermediate sized engines, and Cell 2 is a vertical test stand.

A Test Operations Building (TOB) houses the TCC (Test Control Center) for each of the three test facilities in the E Complex, test control and data acquisition equipment, general offices, and a component preparation area. The component preparation area will be used for assembly and maintenance work on test articles. The TOB will provide an environmentally controlled area for electronic equipment as well as an enclosed facility for personnel safety during testing. The TOB is designed to allow expansion of the data collection and evaluation areas into the component preparation area. Signal Conditioning Buildings (SCBs) are located near the test cells and house test control and data acquisition equipment.

3.3.1 E-1 Test Facility [44]

The E-1 Test Facility is a versatile test complex that is available for component development testing of combustion devices, turbopump assemblies, and other rocket engine components. Originally designed to test turbomachinary for the Advanced Launch System (ALS) and then the National Launch System (NLS) Programs, it has been modified several times. The facility has the capacity to deliver high flow rate propellants at high and low pressures. In addition, the facility is equipped to supply a wide range of supporting fluids. Data acquisition and control systems at E-1 have high capacities and are the state-of-the-art in number of channels and data storage.

The facility has three test cells capable of accommodating multiple programs at the same time. The test article structural supports in all three test cells have common interface arrangements. Each test cell is 30 ft. wide by 30 ft. deep by 26 ft. high. A structural blast wall separates the supporting test complex infrastructure from the test cells. Each test cell contains all of the support equipment, service connections, test instrumentation and safety equipment required to perform testing. A 10-ton overhead bridge crane spans all three test cells and a 10-ton crane provides lifting capability to the facility behind the blast wall. The test cells are covered for protection from environmental elements.

28

E1 Test Stand

3.3.1.1   Cell 1

Cell 1 is primarily designed to test pressure-fed LOX/LH$_2$, LOX/RP-1, and hybrid combustion devices up to 750,000 lbf of horizontal thrust (1,500,000 lbf impulse load). Currently, the high pressure LOX feed line and a high pressure LH$_2$ feedline are routed to Cell 1. Future plans call for a high pressure RP-1 feed line and a low pressure RP-1 feedline to be routed to Cell 1. The following table outlines the existing and predicted future commodity supply capabilities for Cell 1.

Commodity	Pressure (psig)	Temperature (°R/°F)	Flow Rate (lbm/sec)	Supply Line (in.)	Existing/Future	
HP LOX	7,700	178 / -281	1,800	12	☑	☐
LP LH$_2$*	250	38 / -422	225	12	☑	☐
HP LH$_2$	7,750	75 / -385	205	12	☑	☐
LP RP-1	300	540 / 80	1,050	12	☐	☑
HP RP-1	7,800	540 / 80	1,050	12	☐	☑

Note: LP LH2 Facility runline enters at Cell 2, STE piping can route to Cell 1.

3.3.1.2   Cell 2

Cell 2 is primarily designed to test LH_2 and LOX turbopump assemblies, either simultaneously or individually. The test article support structure for Cell 2 is designed for test articles weighing up to 30,000 lbs and generating up to 60,000 lbf thrust (120,000 lbf impulse) at angles up to 10° above horizontal. The following table outlines the existing and future commodity supply capabilities for Cell 2.

Commodity	Pressure (psig)	Temperature (°R/°F)	Flow Rate (lbm/sec)	Supply Line (in.)	Existing/Future	
HP LOX	7,765	178 / -281	165	3	☑	☐
HP LOX	7,765	178 / -281	12	1	☑	☐
LP LOX	295	165 / -294	1,220	12	☑	☐
HP LH₂	7,750	75 / -385	205	8	☑	☐
HP LH₂	7,750	75 / -385	2	1	☑	☐
LP LH₂	250	38 / -422	225	12	☑	☐
LP RP-1*	300	540 / 80	1,050	12	☐	☑
HP RP-1	7,800	540 / 80	55	4	☐	☑

* Note: LP RP-1 Facility runline enters at Cell 1, STE piping can route to Cell 2.

3.3.1.3   Cell 3

Cell 3 is primarily designed to test LOX-rich turbopump assemblies. The test article support structure for Cell 3 is designed for test articles weighing up to 30,000 lbs and generating up to 60,000 lbf thrust (120,000 lbf impulse) at angles up to 10° above horizontal. The following table outlines the existing commodity supply capabilities for Cell 3.

Commodity	Pressure (psig)	Temperature (°R/°F)	Flow Rate (lbm/sec)	Supply Line (in.)	Existing/Future	
HP LOX	7,765	180 / -280	1,002*	2 @ 8	☑	☐
HP LOX	7,765	180 / -280	1.5	1	☑	☐
LP LOX	295	165 / -296	1,220	12	☑	☐
HP LH₂	7,750	61 / -399	30	3	☑	☐
HP LH₂	7,750	61 / -399	0.5	1	☑	☐

* Note: Each 8" supply line provides up to 501 lbm/s.

3.3.1.4 Commodity Storage and Capabilities

The following table contains a complete listing of E-1 facility pressure vessels.

Vessel Number	Vessel Commodity	MAWP (psig)	Volume
V-10A03-GH	GH$_2$	15,000	625 ft^3
V-10A04-GH	GH$_2$	15,000	625 ft^3
V-10A05-GH	GH$_2$	15,000	625 ft^3
V-10F02-GN	GN$_2$	15,000	625 ft^3
V-10F03-GN	GN$_2$	15,000	625 ft^3
V-10A06-GN	GN$_2$	4,500	1,500 ft^3
V-10A10-HE	HE	4,500	1,500 ft^3
V-10A14-LH	LH$_2$	150	5,513 gal
V-10A15-LH	LH$_2$	33	50,000 gal
V-10A17-LH	LH$_2$	385	15,140 gal
V-10F01-LH	LH$_2$	8500	5,000 gal
V-10A20-LN	LN$_2$	165	28,000 gal
V-10A13-LO	LOX	9,000	2,600 gal
V-10A22-LO	LOX	0	28,000 gal
V-10A23-LO	LOX	165	28,000 gal
V-10A24-LO	LOX	400	11,240 gal

3.3.1.4.1 Liquid Oxygen (LOX) (1-6)

A 28,000 gallon LOX vessel provides storage capacity for the 2,600 gallon high pressure LOX run tank and the 11,240 gallon low pressure LOX run tank. A second 28,000 gallon vessel serves as a LOX catch tank with a test article discharge line running to it from the test cells. The following table provides the LOX storage capacities available at E-1.

Vessel Number	Description	Max Pressure (psig)	Water Volume (gal)	Clean Level	Existing / Future	
V-10A13-LO	HP Run Tank	9,000	2,600	1X	☑	☐
V-10A24-LO	LP Run Tank	400	11,240	1X	☑	☐
V-10A23-LO	Storage	165	28,000	1	☑	☐
V-10A22-LO	Catch Tank	0	28,000	2A	☑	☐

3.3.1.4.2   Liquid Hydrogen (LH₂) (18-24)

A 50,000 gallon spherical tank provides liquid hydrogen storage at E-1. Run tank capacities consist of a 5,000 gallon high pressure LH₂ run tank and a 15,140 gallon low pressure LH₂ run tank. In addition, a 5,513 gallon LH₂ tank is used in generating high pressure gaseous hydrogen as discussed in section 4.3. A high pressure LH₂ flarestack located at E-1 is capable of burning LH₂ at test article discharge flow rates up to 200 lbm/sec at 3,575 psig. The following table provides the LH₂ storage capacities available at E-1.

Vessel Number	Description	Max Pressure (psig)	Water Volume (gal)	Clean Level	Existing / Future	
V-10F01-LH	HP Run Tank	8,500	5,000	2X	☑	☐
V-10A17-LH	LP Run Tank	385	15,140	2X	☑	☐
V-10A15-LH	Storage	33	50,000	2	☑	☐
V-10A14-LH	Day Tank	150	5,513	2	☑	☐

3.3.1.4.3   Gaseous Hydrogen (GH₂)

The E-1 facility has a total of 1,875 ft³ ultra-high pressure (UHP) GH₂ storage capacity. An additional 1,250 ft³ of UHP GH₂ storage capacity is planned. The ultra-high pressure GH₂ is used to pressurize the LH₂ run tanks and to temperature condition LH₂ supplied to test cells 2 & 3. Site wide GH₂ supply is available at E-1 up to 3,000 psig. For higher pressures, LH₂ from the 5,513 gallon day tank is pressurized using an ultra-high pressure pump and then vaporized. A low pressure GH₂ flarestack located at E-1 is capable of burning GH₂ at test article discharge flow rates up to 200 lbm/sec at 600 psig. The following table provides the existing and future GH₂ storage capacities available at E-1.

Vessel Number	Description	Max Pressure (psig)	Water Volume (ft³)	Clean Level	Existing / Future	
V-10A03-GH	UHP Storage	15,000	625	2	☑	☐
V-10A04-GH	UHP Storage	15,000	625	2	☑	☐
V-10A05-GH	UHP Storage	15,000	625	2	☑	☐
TBD	UHP Storage	15,000	625	2	☐	☑
TBD	UHP Storage	15,000	625	2	☐	☑

3.3.1.4.4   RP-1

Future plans call for a low pressure and high pressure RP-1 systems at E-1. Also available will be a 20,000 gallon RP-1 storage tank and a 2,000 gallon RP-1 catch tank with a RP-1 discharge line routed to it from the test cells. The following table provides the proposed RP-1 storage capacities to be available at E-1.

Vessel Number	Description	Max Pressure (psig)	Water Volume (gal)	Clean Level	Existing / Future	
V-464-RP	LP Run Tank	400	7,000	2	☐	☑
V-471-RP	HP Run Tank	9,000	2,600	2	☐	☑
V-465-RP	Storage	0.5	20,000	3	☐	☑
V-TBD-RP	Catch Tank	50	2,000	3	☐	☑

3.3.1.4.5   Liquid Nitrogen (LN₂) (36)

A 28,000 gallon vessel provides liquid nitrogen storage at E-1. The LN_2 is used in generating high pressure gaseous nitrogen as discussed in section 3.3.1.4.6 and can also be used in "cold shock" and other activation activities. As necessary, the LOX discharge line and 28,000 gallon LOX catch tank described in section 3.3.1.4.1 can be used in LN_2 service. The following table provides the LN_2 storage capacity available at E-1.

Vessel Number	Description	Max Pressure (psig)	Water Volume (gal)	Clean Level	Existing / Future	
V-10A20-LN	Storage	165	28,000	1	☑	☐

3.3.1.4.6   Gaseous Nitrogen (GN₂) (37, 38)

The E-1 facility currently has 1,250 ft^3 of ultra-high pressure GN_2 storage capacity and 1,375 ft^3 of high pressure GN_2 storage capacity. An additional 625 ft^3 of UHP GN_2 storage capacity is planned. The ultra-high pressure GN_2 is used to pressurize the high pressure LOX run tank and the high pressure GN_2 is used to pressurize the low pressure LOX run tank. Site wide GN_2 supply is available at E-1 up to 4,500 psig. For higher pressures, LN_2 from the 28,000 gallon storage tank is pressurized using an ultra-high pressure pump and then vaporized. The table on the next page provides the GN_2 storage capacities available at E-1.

Vessel Number	Description	Max Pressure (psig)	Water Volume (ft^3)	Clean Level	Existing / Future	
V-10F02-GN	UHP Storage	15,000	625	1	☑	☐
V-10F03-GN	UHP Storage	15,000	625	1	☑	☐
V-10F04-GN	UHP Storage	15,000	625	1	☐	☑
TBD	UHP Storage	15,000	625	1	☐	☑
TBD	UHP Storage	15,000	625	1	☐	☑
V-10A06-GN	HP Storage	4,500	1,500	1	☑	☐

3.3.1.4.7 Gaseous Helium (GHe)

The E-1 facility has 1,375 ft^3 of high pressure GHe storage capacity. Site wide GHe supply is available at E-1 up to 4,400 psig. In addition, GHe can also be supplied by tube bank trailers. The following table provides the GHe storage capacity available at E-1.

Vessel Number	Description	Max Pressure (psig)	Water Volume (ft^3)	Clean Level	Existing / Future	
V-10A10-HE	HP Storage	4,500	1,500	1	☑	☐

3.3.1.5 Electrical System Capabilities

The capabilities described below are for each of three test cells when they are completed.

3.3.1.5.1 Control System

The E-1 control system consists of a number (currently ten) of Programmable Logic Controller (PLC) cabinets that are assigned specific tasks in the control and operation of the facility. A network that allows them to share information or tasks connects them. I/O and control tasks can be reassigned to meet specific performance goals of a particular test program.

Facility control sub-systems provide for the transfer of propellants to storage and run tanks, operation of pumps and vaporizers, and pressurization of run tanks. Generally these subsystems are not actively involved during the actual test. Facility alarms for out-of-tolerance conditions are provided. The facility controls provide a real-time display of operations in a process diagram format.

Other sub-systems provide control over systems that are used for real time control during tests. While many control functions, such as run tank pressurization, are shared by several or all test cells, one PLC is reserved per test cell for test article or test program specific tasks. Redline capability is provided for all of these sub-systems. An additional dedicated redline PLC sub-system is provided for monitoring test specific or test article instrumentation redlines. The system provides the ability to connect up to 80 redline monitoring and shutdown (cutoff) measurements. The PLCs have a nominal loop response (ladder logic scan time) of 30 ms, which does not include analog I/O, instrumentation, or valve delays.

A hardwired emergency abort system is provided to manually override the control system and safely shut down the facility and associated test article systems in a predetermined sequence.

3.3.1.5.2   Instrumentation

Facility instrumentation is installed for real-time display of facility processes and data recording. Instrumentation provided for the test article will be determined with each customer. The facility also provides the ability to display real-time test article measurements. Facility instruments not used during the test, such as those in the storage and transfer areas have a typical response of 250 ms. Instruments that provide test article response information, time critical control functions, or are located at facility to test article interfaces range from 5 to 150 ms response depending on the specific instrument.

Up to 320 channels of general-purpose signal conditioning are available to excite, amplify, and filter a variety of sensors in each test cell. Up to 80 channels of standard thermocouple signal conditioning is also available in each test cell. In addition, other special signal conditioning, such as charge amps, band-pass filters, RMS, and frequency to DC converters are available.

3.3.1.5.3   Low Speed Data Acquisition System

Each test cell (test article & STE) has a dedicated Low Speed Data Acquisition System (LSDAS). There is also one dedicated to the test facility. Each LSDAS can provide up to 500 analog channels and 300 discrete channels with a total throughput of 500K sps of 16 bit data. Each test cell LSDAS has the capability to patch additional channels from an adjacent cell's LSDAS. IRIG B time is recorded on the LSDAS for time correlation between systems. Data is recorded remotely in the Test Control Center (TCC).

The LSDAS uncertainty is expected to be better than ±0.15%, excluding instrumentation and associated cabling. The LSDAS provides a real-time display with an updating rate typically set at 3 times per second. Real-time calculated value capability is available.

3.3.1.5.4 High Speed Data Acquisition System

The High Speed DAS (HSDAS) has three high speed 16 bit digitizers and recorders per test cell. Each digitizer can be configured as follows:

- 62 channels at 50 KSPS (22.5 KHz)
- 32 channels at 100 KSPS (45.5 KHz)
- 16 channels at 200 KSPS (80 KHz)

Typically each digitizer is configured for 32 channels at 100 KSPS per channel. The data is recorded digitally on Super VHS tape along with IRIG B time for correlation. Event channels are also available.

3.3.1.5.5 Data Processing

Data processing is provided for both the LSDAS and the HSDAS. The LSDAS data is converted to Engineering Units and processed into the E-complex standard file format. The LSDAS data file can be provided in Winplot, binary, or an ASCII format. A measurement and calculated values plotting program is available. A quicklook program is also available for on-line analysis. Data backup is provided on DAT or CD-ROM.

The HSDAS data is generally filtered, subsampled, and processed to Fast Fourier Transforms (FFT) for frequency analysis. A quicklook dynamic analysis program is available for data reports and quicklook. FFT data format is the same as SSME. Subsampled time domain data can be provided in Winplot format. Data backup is provided on CD-ROM for FFT data and SVHS for raw dynamic data.

Off-site data transfer is available for both systems through the Internet.

3.3.1.5.6 Power

AC - The facility power system provides single and three-phase power at 480VAC, 277VAC, 208VAC, 220VAC and 120VAC. Uninterruptible Power Supply (UPS) systems provide 220VAC and 110 VAC power to test critical systems.

DC - 28V DC is provided for control system I/O.

3.3.1.5.7 Video Systems

A low speed video system is available for test article and facility monitoring. Video displays/recorders are provided in the TCC.

A high-speed (HS) video system is used for test article monitoring and recording. Currently two HS video cameras are provided routinely with others provided on an as-available basis. The system is capable of providing up to 500 images per second, available for immediate playback after the test is complete.

IRIG B time is recorded on all systems including video for time correlation.

3.3.1.6   Ancillary Facility Systems

3.3.1.6.1   Deluge Water

The E-Complex shares a 4,000 GPM deluge water system for the purpose of providing deluge cooling water and limiting damage in the event of a fire in a test cell or propellant storage/handling area. Since the system is shared, test activities requiring full deluge capability must be coordinated between facilities. The system covers the following areas: all test cells, all active LOX vessels, all active LH_2 vessels, all active tanker fill headers, and the hydraulics skid. All areas can be remotely operated from the control room. In addition, the deluge nozzles covering the tanker fill headers can be locally operated.

3.3.1.6.2   Plume Impingement Area

A plume impingement area extending southward from Cell 1 is engineered to minimize the effects of heat and acoustic loads generated by a combustion device plume. Water spray nozzles supplied by the deluge water system also cover the plume impingement area.

3.3.1.6.3   Hydraulics

The facility is equipped with a hydraulic system for actuating facility and special test equipment hydraulic valves. The hydraulic system is capable of supplying 180 GPM at 3,000 psig. A smaller jockey pump is capable of supplying 20 GPM at 3,000 psig for low load conditions.

3.3.1.6.4   Communications

The E1 facility is equipped with headset communication boxes in numerous locations to allow voice communication between the Test Control Center and the test facility during operations.

3.3.2 E-2 Test Facility

The E-2 test facility, formerly known as the High Heat Flux Facility (HHFF), was originally constructed to support materials development testing for the National Aerospace Plane (NASP) by subjecting special test articles to extreme temperature conditions. Recent test projects at E-2 have included composite cryogenic tanks, turbopumps, and preburners. The facility is now being modified to support advanced component and engine development projects. The facility has two test cells capable of accommodating multiple programs at the same time.

Cell 1 is an open air covered test cell equipped with a 2 ton Jib crane. A structural blast wall separates the supporting test complex infrastructure from the test cell. Gaseous propellant run, storage and transfer systems are available. A Pebble Bed Heater provides a hot-gas capability. The facility also includes a Signal Conditioning Building (SCB) and an Electrical Equipment Building (EEB). All data acquisition and control aspects are fully functional and are supported by the E-2 TCC, located within the Test Operations Building (TOB).

Cell 2 is capable of supporting tests of complete flight or "flight-like" stages, as well as rocket engines and combustion devices. When the facility upgrades are complete in 2001, Cell 2 will have the capability to provide low pressure LOX and RP-1 propellants to a test article, mounted vertically in the test cell. A facility Thrust Measurement System (TMS) is available to measure test article thrust levels. LOX and RP-1 Flow Measurement Systems (FMS), utilizing turbine flow meters, are available to measure LOX and RP-1 flow rates from facility run tanks.

E-2 Cell 1

3.3.2.1 Cell 1 [45]

Cell 1 is a 20 ft. by 20 ft. covered test cell. It can withstand up to 100,000 lbf of thrust from a combustion device or turbopump. It has the capability of providing the test article with LOX, LN_2, LH_2, RP-1, water, GH_2, GOX, and GN_2. The facility is equipped with an ablative coated concrete pad located east of the test cell and is capable of withstanding the pressure and heat loads generated by a 100,000 lbs thrust class combustion device. The pad is covered by spray nozzles supplied with water from the deluge water system in order to reduce the temperature of the impingement area. The location and size of the concrete pad can accommodate combustion devices mounted at the top of the test article support stand (120 inches from the floor of the cell) and canted downward at up to a 10 degree angle from the horizontal.

3.3.2.1.1   Commodity Storage and Capabilities

LOX, LN$_2$, LH$_2$, RP-1, water, GH$_2$, GOX, and GN$_2$ can be supplied to test articles in E-2 in accordance with the facility pressure vessels as shown in the following table.

Vessel Number	Vessel Commodity	MAWP (psig)	Volume
Propellant Run Systems			
V-10A21-RP	RP-1	9,000	500 gal.
V-14A12-RP	RP-1	6,000	145 gal.
V-14A11-RP	RP-1	1,800	846 gal.
♦V-TBD-RP	RP-1	200	7,000 gal.
V-065-GO	GO$_2$	4,500	1,375 ft^3
V-14A19-LO	LO$_2$	9,300	500 gal.
V-14A10-LO	LO$_2$	185	10,000 gal.
V-14A13-CW	RP-1/H$_2$0	6,000	145 gal.
♦V-TBD-LH	LH$_2$	9,000	1,300 gal
V-456-LH	LH$_2$	4,000	4,500 gal.
V-473-LH	LH$_2$	400	15,000 gal.
X-14X70-GH	Hot GH$_2$	6,600 (400 F)	65 ft$^{3\otimes}$
Pressurization Systems			
V-10A09-GN	GN$_2$	15,000	247 ft^3
V-145-GN	GN$_2$	5,667	1,500 ft^3
V-142-GN	GN$_2$	5,667	1,375 ft^3
V-14A18-GH	GN$_2$/GH$_2$	15,000	625 ft^3
V-264-GH	GH$_2$	6,600	750 ft^3
V-265-GH	GH$_2$	6,600	750 ft^3
Propellant Storage/Dump Systems			
V-14A01-LO	LO$_2$	150	3,000 gal.
♦V-458-LO	LO$_2$	165	28,000 gal.
♦V-TBD-LH	LH$_2$	165	28,000 gal.
♦V-TBD-RP	RP-1	30	28,000 gal.

Table 3.3.2.1.1-1 E-2 Cell 1 Facility Pressure Vessels

System	Max. Tank Opr. Press (psig)	Max. STE Interface Pressure (psig)	Max. Flow Rate (lbm/sec)	Duration for Max. Press/flow rate Case (sec)	Amount of Commodity Used for Max. Press/flow rate Case	Available Commodity with Standard Margins	Press. System Ramp Rate (psig/sec)
UHP LO$_2$	8,000	7,500	275	5	144 gal.	400 gal.	1,000
LP LO$_2$	165	100	275	277	8,000 gal.	8,000 gal.	7.4 (7)
(6)GO$_2$	4,100	3,000	65	100	6,500 lbm	n/a	n/a
LO$_2$ Discharge	n/a	600	275	70	2000 gal.	n/a	n/a
UHP RP-1	8,000	7,500	140	19	400 gal.	400 gal.	700
HP RP-1 (coolant)	5,500	5,000	60	12	107 gal.	116 gal.	4,600
MP RP-1	1,600	1,400	140	30	623 gal.	677 gal.	285
♦LP RP-1	360	320	140	265	5,500 gal.	5,500 gal.	TBD
♦RP-1 Discharge	30	600	140	1200	25,200 gal.	n/a	n/a
♦UHP LH$_2$	8,000	7,500	55	8	544 gal.	664 gal.	500
HP LH$_2$	3,600	3,200	55	30	2,240 gal.	2,240 gal.	300
LP LH$_2$	360	320	55	112	10,500 gal.	10,500 gal	10
(4)GH$_2$	6,000	3,000	19	35	675 lbm	n/a	n/a
(3)Hot GH$_2$	6,000	4,000 @ 1300 °F	10	40	400	n/a	n/a
12" LH$_2$ Flare-stack	n/a	3,000	15	n/a	n/a	n/a	n/a
24" LH$_2$ Flare-stack	n/a	1,200	40	n/a	n/a	n/a	n/a
42" GH$_2$ Flare-stack	n/a	3,000	81	n/a	n/a	n/a	n/a
HP H$_2$0 (coolant)	5,500	5,000	75	12	106 gal.	116 gal.	4,600
(5)GN$_2$	6,000	3,000	70	95	6,630 lbm	n/a	n/a
TEA/ TEB	5,500	5,000	1.0	6	1.0 gal.	1.5 gal.	n/a

Table 3.3.2.1.1-2   E2 Cell 1 Propellant/Coolant/Turbine Drive/Discharge Systems Capabilities

Table 3.3.2.1.1-1 and Table 3.3.2.1.1-2 Notes:

(1) Pressurization system ramp rate is a static ramp rate of the run tank ullage with no flow out of the tank. These are nominal predicted values. These ramp rates do not include valve actuation times, which could be a significant factor on the high pressure vessel pressurization rates. All cases assume 10% ullage volume and pressurization from 0 to MAWP of vessel.

(2) "Amount of commodity available using standard margins" column includes standard reductions in available propellant volume for ullage (10%), residual/heel (10%), and, for LOX, boil-off/bleed (10%), which gives a total volume reduction of 20% for RP-1, H$_2$O; 30% for LOX, LN$_2$, LH$_2$.

40

(3)   The hot hydrogen system temperature capability in terms of test duration at the specific temperature will be determined during activation. The design target is 100 seconds at 4.0 lbm/sec at 2,500 psia at 750 °F ± 100 °F

(4)   Case with GH_2 bottles (V-264-GH and V-265-GH) used in GH_2 service for either turbine drive or combustion device application, blowing down from 6,000 psig to 3,500 psig. Maximum flowrate provided duration illustration only.

(5)   Case with GH_2 bottles (V-264-GH and V-265-GH) used in GN_2 service for turbine drive application, blowing bottle down from 6,000 psig to 3,500 psig.

(6)   Case with GOX bottle (V-065-GO) used in combustion device application, blowing down from 4,100 psig to 3,500 psig. maximum flowrate based on maximum safe GOX velocity of 125 ft/sec.

(7)   Assumes 3,000 psi pressurant bottle pressure. Ramp rate is limited by regulator (not piping losses) therefore changing pressurization regulators from planned Grove 301B 0.5" orifice to Grove 301B 1" balanced valve would increase this to 18.5 psig/sec.

⊗   Vessel volume minus refractory brick and pebbles (gas volume) is 27.5 ft^3

♦   Future capability

3.3.2.1.1.1   Liquid Oxygen (LOX)

Cell 1 contains a 6,000 psi/6 GPM LOX pump (P-14Q36-LO) for the purpose of pressurizing the GOX vessel (V-065-GO) to 4,500 psi via a 6,000 psi vaporizer (X-14Q46-LO). This pump is supplied by LOX vessel V-14A01-LO. The GOX vessel can be charged from 0 psig to 4,500 psig in 15 hours.

3.3.2.1.1.2   Liquid Hydrogen (LH$_2$)

Cell 1 has a flarestack capable of burning test article LH_2 discharges as specified in the earlier table(s).

3.3.2.1.1.3   Gaseous Hydrogen (GH$_2$)

Cell 1 has an intensifier (level 2 clean) for the purpose of pressurizing the GH_2 vessels (V-264-GH, V-265-GH, and V-14A18-GH) up to 6,600 psi with either GN_2 or GH_2. The intensifier can charge one vessel from site-wide average 3,000 psig to 6,600 psig in approximately 8 hours.

The Cell 1 Ultra High Pressure vessels for GH_2/GN_2 (V-14A18-GH, 15,000 psi/625 ft^3 level 2 clean) are rechargeable from the site-wide average 3,000 psig to 15,000 psig in 12 hours. This is accomplished using the E-1 facility UHP system and a cross-country line between E-1 and E-2.

Cell 1 is equipped with a flarestack capable of burning test article GH_2 discharges as specified in the earlier table(s).

3.3.2.1.1.4 RP-1

Cell 1 has an RP-1 discharge line routed to a waste tank at the north side of the test cell blast wall. This system has the capabilities specified in the earlier table(s).

3.3.2.1.1.5 Liquid Nitrogen (LN₂)

LN_2 is not currently available at E2 Cell 1.

3.3.2.1.1.6 Gaseous Nitrogen (GN₂)

The E-2 Ultra High Pressure vessels for GH_2/GN_2 (V-14A18-GH, 15,000 psi/625 ft^3 level 2 clean) and GN_2 (V-10A09-GN, 15,000 psi/247 ft^3 level 1 clean) are rechargeable from the site-wide average 3,000 psig to 15,000 psig in 12 hours. This is accomplished using the E-1 facility UHP system and a cross-country line between E-1 and E-2.

A 6,000 psi/6 GPM LOX pump (P-14Q36-LO), supplied by LOX vessel V-14A01-LO, can be used in LN_2 service to charge the GN_2 vessels (V-142-GN and V-145-GN) from site-wide average 3,000 psig to 5,667 psig in 9 hours or vessel V-10A09-GN to 6,000 psig in approximately 3 hours.

3.3.2.1.1.7 Gaseous Helium (GHe)

Site wide GHe supply is available at E-2 up to 4,400 psig.

3.3.2.1.2 Electrical System Capabilities

The following sections will briefly describe the control system, instrumentation, Low Speed and High Speed DAS systems, data processing, power and video systems.

3.3.2.1.2.1 Control System

The Test Control Center for Cell 1 is located in Bldg. 4010. The facility control system handles the transfer of propellants to storage and run tanks, operation of pumps and vaporizers, and pressurization of run tanks. Facility alarms for out of tolerance conditions are provided. The facility controls provide a real-time display of operations in a process format.

A test article control system contains a general automatic sequencer that can be configured for a test article specific test sequence. The system provides (via patching from the respective DAS signal conditioners) the ability to connect sixty-four (64) analog redline monitoring and shutdown (cutoff) measurements to the Programmable Logic Controller (PLC). Of these 64 channels, up to sixteen (16) of these measurements may be High Speed Instruments, such as accelerometers or speed sensors.

A PLC is also used for run tank pressurization, temperature ramping, and valve sequencing. This PLC can be used for analog outputs of up to sixty-four (64) STE and/or Test Article devices of which thirty-two (32) can control hydraulic servo valves.

A hardwired Emergency Shutdown System is provided that manually overrides the control systems and shuts down the facility and associated test article systems in a predetermined timed sequence.

3.3.2.1.2.2 Instrumentation

Facility instrumentation is installed for real-time display of facility processes and data recording. Instrumentation provided for the test article will be determined with each customer. The facility provides the ability to display real-time article measurements. Measurements that provide test article response information, time critical control functions, or are located at facility to test article interfaces have a 5 ms response, except sheathed thermocouples which have a 250 ms response and RTDs which have a 500 ms response.

3.3.2.1.2.3 Low Speed Data Acquisition System

The Low Speed Data Acquisition System (LSDAS) provides 360 channels with a total throughput of 200 KSPS of 16 bit data. Test Article and Control system data are recorded on the LSDAS. IRIG B is also recorded on the LSDAS for time correlation between systems. The DAS end-to-end uncertainty is $\pm0.15\%$, excluding instrument and associated cabling. The LSDAS provides a real-time tabular display with an updating rate of 3 sps per measurement and redundant data recording. A limited number of real-time calculated values are available, depending on complexity and available LSDAS overhead.

3.3.2.1.2.4 High Speed Data Acquisition System

The High Speed DAS (HSDAS) throughput is 50 KSPS per channel at 16 bits for 64 channels. The data is recorded on Super VHS tape along with IRIG B for time correlation. Discrete event channels are also available.

3.3.2.1.2.5 Data Processing

Data processing is provided for both the LSDAS and the HSDAS. The LSDAS data is converted to Engineering Units and processed into the E-complex standard file format. The LSDAS data file can be provided in an ASCII format. A measurement and calculated values plotting program is available. A quicklook program is also available for on-line analysis. Data backup is provided on DAT or CD ROM.

The HSDAS data is generally filtered and subsampled to 5KHz and processed to FFT for frequency analysis. A quicklook dynamic analysis program is available for data reports and quicklook. Data backup is provided on CD ROM for FFT data and SVHS for raw dynamic data.

Post test, off-site data transfer is available for both data systems through the Internet. Data security is provided through password protection.

3.3.2.1.2.6 Power

AC - The facility power system shall provide 480VAC, 220VAC, 208VAC, and 110VAC; single and three-phase.

DC - The facility power system shall provide 28V DC for valve control.

3.3.2.1.2.7 Video Systems

A standard VHS video system (30 frames/sec) is available for test article and facility monitoring. Video displays/recorders are provided in the test control room.

Cell 1 has a three position high speed video system used for test article monitoring and recording. Currently video cameras are provided on an as available basis, with video mounts and cabling as part of the E-2 facility. Future upgrades will provide dedicated high speed video cameras.

3.3.2.1.3 Ancillary Facility Systems

The following sections describe the test article support stand, hydraulics system, plume impingement area and deflector, deluge system and communications.

3.3.2.1.3.1 Test Article Support Stand

The E-2 Cell 1 test article support stand is designed to support a test article weighing up to 3,000 lbs and generating up to 100,000 lbs of thrust (200,000 lbs impulse load). The orientation of the support stand has the test article plume directed eastward from the test cell. The support stand is designed such that the test article centerline can be mounted from 60 to 120 inches off the test cell floor (the test cell floor is 30 inches below the grating). The test stand is capable of mounting a combustion device at up to a 10 degree downward angle from the horizontal and is capable of taking side loads of up to 15,000 lbs. The stand is equipped with an axial direction thrust measurement system with accuracy of 0.5 percent of load over the thrust range of 10,000 to 100,000 lbs.

3.3.2.1.3.2 Hydraulics System

The facility includes a hydraulic system for actuating facility and STE hydraulic valves. The system at Cell 1 is capable of generating 3,000 psig and 40 GPM with a 150 gallon hydraulic fluid reservoir.

3.3.2.1.3.3   TEA/TEB Ignition System

Cell 1 is equipped with a Triethyl-Aluminum (TEA) and Triethyl-Borane (TEB) combustion device ignition system. This system has the capabilities as specified in the earlier table(s).

3.3.2.1.3.4   Deluge System

The facility has a 4,000 GPM deluge system for the purpose of limiting damage in the event of a test stand fire. The deluge system water supply is only available when E-1 is not involved in a test requiring deluge support.

The Cell 1 Deluge System covers the following areas: Test cell, all active LOX vessels, all active LH_2 vessels, all active RP-1 vessels, all active GOX vessels, the Pebble Bed Heater, all active tanker fill headers, and the hydraulics skid. The deluge nozzles covering the tanker fill headers are controllable both remotely and locally whereas all other areas are only remotely operated from the control room.

3.3.2.1.3.5   Communications

The facility communications system has the capability to provide communications between the Test Control Center and the test cell.

3.3.2.1.5.6   Fire and Gas Detect System

The E2 Test Complex employs a fire and gas detection system with a display within the E2 Test Control Center (TCC) that is independent of the Facility and Test Article Control System. This fire and gas detect system provides operators within the (TCC) visibility into severity and location of fires and Gaseous Hydrogen leaks.

3.3.2.1.5.6   Pebble Bed Heater

A hot hydrogen capability is provided by a Pebble Bed Heater with capability as shown in Table 3.3.2.1.1-1. This system has been installed and partially activated, awaiting a customer need.

3.3.2.2   Cell 2 [46]

Currently, the E-2 Cell 2 Test Facility is capable of testing complete flight stages or "flight-like" test article stages up to 120,000 lbf thrust in a vertical orientation.

E 2 Cell 2

The vertical test cell, which runs from the rolling deck platform to the top of the test cell structure, measures 22' 4" x 22' 4" x 58' 10". The diamond-shaped thrust takeout structure reaction beams provide a 15 ft x 15 ft area to accommodate test articles. Larger test articles, up to about 22 ft in diameter, can be accommodated by making structural modifications to the test cell. Access stairs are provided at both the north and south sides of the test stand.

A facility Thrust Measurement System (TMS) is available, and facility FMS, using turbine flowmeters for measuring LOX and RP-1 flow rates from the facility run tanks, will be available as part of the facility upgrades.

3.3.2.2.1   Commodity Storage and Capabilities

Facility pressure vessels are listed in Table 1, along with their Maximum Allowable Working Pressures (MAWP), volumes, and cleanliness levels. Table 2 summarizes the capabilities of the LOX and RP-1 run systems, including operating pressures, maximum flow rates, and run durations. The values provided in Table 2 are nominal predicted values. Actual values will be determined during facility activation prior to test. The facility currently provides 2" LOX and RP-1 transfer lines from the facility storage tanks to the test cell for test article use.

Table 1: E2 Cell 2 Facility Pressure Vessels

Vessel Number	Vessel Commodity	MAWP (psig)	Volume (gal)	Clean Level[1]	Existing/Future	
Run Tanks						
V-103-LO	LOX	120	5,500	1XX	☐	☑
V-196-RP	RP-1	125	5,500	2X	☐	☑
Storage Tanks						
V-14A1200-LO	LOX	250	13,000	1	☑	☐
V-14A1210-RP	RP-1	14" H_2O	15,000	3	☑	☐
Run Tank Pressurization Accumulators						
LOX: ACC-14A1785-GN	GN_2	2,250	1,250	1XX	☐	☑
RP-1: ACC-14A1873-GN	GN_2	2,250	500	2X	☐	☑

1.) As per SSC STD 79-001

Table 2: E2 Cell 2 Propellant/Coolant/Turbine Drive/Discharge Systems Capabilities

System	Max. Tank Press. (psig)	Max. STE Interface Pressure[1] (psig)	Max. Flow Rate (lbm/sec)	Duration (at max. press. & flow rate) (sec)	Available Propellant (std. margins)[2] (gal)	Max. Press./Vent Rate[3] (psig/sec)	Run Line Diam. (in)	Exist./Future	
LOX Run	120	70	300	122	3,850	1.0	6	☐	☑
RP-1 Run	125	64	140	212	4,400	1.0	4	☐	☑

1.) Calculated estimate for maximum propellant flow rates
2.) Includes standard reductions in available propellant volume for ullage (10%), residual/heel (10%), and, for LOX, boil-off/bleed (10%), which gives a total volume reduction of 20% for RP-1 and 30% for LOX
3.) For maximum propellant flow conditions

3.3.2.2.1.1 Liquid Oxygen (LOX)

The maximum operating pressure is 120 psig for the LOX facility run system. The pressurization system will be capable of pressure ramping up or down at 1.0 psi/sec (at maximum propellant flow rates) as part of the planned facility upgrades. Similar pressurization ramp rates will be available for lower propellant flow rates.

The Cell 2 LOX facility run line is equipped with a turbine flowmeter based FMS. The LOX FMS includes flow straightener/conditioner spool pieces upstream and downstream of the turbine flowmeter. The LOX FMS provides a flow measurement accuracy of 0.5% over a flow rate range of 300 - 3,000 GPM (47 - 470 lbm/sec) with a transient response of 50 ms (i.e., to a flow rate step increase of 10% of maximum flow).

The facility includes a 4 in. LOX dump/bleed line to facilitate chill down and draining of the test article. The line is routed from the test cell to the E-2 Cell 2 LOX dump system, which terminates in a LOX dump pond located at the northeast corner of the E-2 Cell 2 test area.

3.3.2.2.1.2 Liquid Hydrogen (LH$_2$)

LH$_2$ is not currently available at E-2 Cell 2.

3.3.2.2.1.3 Gaseous Hydrogen (GH$_2$)

GH$_2$ is not currently available at E-2 Cell 2.

3.3.2.2.1.4 RP-1

The maximum operating pressure is 125 psig for the RP-1 facility run system.

The RP-1 facility run line is equipped with a turbine flowmeter based FMS. The RP-1 FMS includes flow straightener/conditioner spool pieces upstream and downstream of the turbine flowmeter. The RP-1 FMS provides a flow measurement accuracy of 0.5% of reading over a flow rate range of 125-1,250 GPM (14.1-141 lbm/sec) with a transient response of 50 ms (i.e., to a flow rate step increase of 10% of maximum flow).

A 4 in. RP-1 dump/bleed line is also provided to facilitate test article draining. The line is routed from the test cell to the Cell 2 RP-1 storage tank on the southwest corner of the Cell 2 test area.

3.3.2.2.1.5 Liquid Nitrogen (LN$_2$)

LN$_2$ is not currently available at E-2 Cell 2.

3.3.2.2.1.6 Gaseous Nitrogen (GN$_2$)

E-2 Cell 2 uses the site-wide GN$_2$ system as a source for propellant run tank pressurization and for facility system and test article purges. The 4,100 psig (maximum) site-wide pressure is regulated down to meet lower pressure facility and test article operational requirements. GN$_2$ is available at LOX compatible cleanliness levels (SSC Level 1XX).

3.3.2.2.1.7 Gaseous Helium (GHe)

Site-wide helium (at 4,400 psig, maximum) is also available for pressurization and purging. GHe is available at LOX compatible cleanliness levels (SSC Level 1XX).

3.3.2.2.2   Electrical System Capabilities

3.3.2.2.2.1   Control System

The E-2 Cell 2 Test Control Center is co-located with the E-2 Cell 1 Test Control Center. It hosts the facility and test article control systems, the data acquisition systems (DAS) user interface, and the test stand video monitoring system. The layout of the E-2 Cell 2 Test Control Center is illustrated in Figure 3.3.2.2.2.1-1.

The facility control system handles the transfer of propellants to storage and run tanks, operation of transfer pumps, and pressurization of run tanks. It provides a real-time display of operations in a process format. Facility blueline and redline alarms for out of tolerance conditions are provided.

Figure 3.3.2.2.2.1-1 E-2 Cell 2 Test Control Center Layout

The test article control system contains a general automatic sequencer that can be configured for test article specific test sequences. The system provides (via patching from the respective DAS signal conditioners) the ability to connect 48 analog redline monitoring and shutdown (cutoff) measurements to the Programmable Logic Controller (PLC).

A PLC is also used for run tank pressurization and valve sequencing. The closed-loop control system is also handled by a PLC.

A hardwired Emergency Shutdown System is provided that manually overrides the control systems and safely shuts down the facility and associated test article systems in a predetermined timed sequence.

3.3.2.2.2.2   Instrumentation

Facility instrumentation is installed for real-time display of facility processes and data recording. Instrumentation provided for the test article will be determined through discussions with each customer. The facility is capable of providing real-time data system test article measurement displays both locally and to off-site locations.

3.3.2.2.2.3   Low Speed Data Acquisition System

In addition to recording facility data, the Low Speed Data Acquisition System (LSDAS) provides 256 total channels for test article and Special Test Equipment (STE) use. These channels include 160 programmable full bridge amplifiers, 60 filter amplifiers (for a total of 220 hardwired channels), and 36 raw input channels. Total throughput is 200 sps of 16 bit data for each channel. Frequency-to-DC converters are available for special applications. Test article and control system data are recorded on the LSDAS. There are 960 discrete and 384 analog LSDAS channels available strictly to record control parameters. The Inter-Range Instrumentation Group (IRIG) B time standard is also recorded on the LSDAS for time correlation between systems. The DAS end-to-end uncertainty is ±0.15%, excluding instrument and associated cabling.

The LSDAS provides a real-time tabular or graphical display of data system measurements. These displays can be transmitted to off-site customer locations. A limited number of real-time calculated values are available, depending on complexity and available LSDAS overhead.

3.3.2.2.2.4   High Speed Data Acquisition System

The High Speed DAS (HSDAS) provides 124 channels for customer use, plus 4 additional channels reserved for IRIG-B timing. Channel throughput is 100 KSPS per channel at 16 bits with a 45 KHz bandwidth. The data is recorded on Super VHS tape along with IRIG B for time correlation. Discrete event channels are also available.

3.3.2.2.2.5   Data Processing

Data processing is provided for both the LSDAS and the HSDAS. The LSDAS data is converted to Engineering Units and processed into the E-complex standard WinPlot file format. WinPlot is a NASA-developed program for plotting measurement and calculated values. A quick look program is also available for on-line analysis. Data backup is provided on digital audiotape or CD ROM formats.

The HSDAS data is available in both time domain and frequency domain formats. Digital filtering algorithms are available for processing. Data backup is provided on CD ROMs for Fast Fourier Transform data and on Super VHS tape for raw dynamic data.

Post test, off-site data transfer is available for both data systems through the Internet. Data security is provided through password protection. Data encryption is also available.

3.3.2.2.2.6 Power

The facility power system provides 480/277 VAC three-phase and 240/120 VAC single-phase power. The 240/120 VAC, supplied from a battery-backed Uninterruptible Power Source (UPS), is supplied to the DAS and control system. 240/120 VAC is also available for test article use.

The facility power system provides 28 VDC for valve control and 24 VDC for instrumentation transmitters. DC power is sourced from the facility UPS. 28 VDC is also available for test article use.

3.3.2.2.2.7 Video Systems

A low speed video system (30 frames/second) is available for test article and facility monitoring. A high speed video system, capable of up to 500 frames/sec, enables 360° viewing of the test article using three high speed cameras with pan/tilt capability placed 120° apart. Video displays and recorders (operating at 30 frames/second) are provided in the Test Control Center. Video tape recorders for high speed video recording are located at the test stand.

3.3.2.2.3 Additional Facility Capabilities

Additional E-2 Cell 2 facility capabilities are described in the sections that follow.

3.3.2.2.3.1 Test Article Support Structure

The E-2 Cell 2 facility is capable of supporting test articles, including complete flight or flight-like stages, generating up to 120,000 lbf thrust (240,000 lbf impulse load) in a vertical orientation. The thrust bearing capability of the facility is currently limited by the load bearing capacity of the flame deflector. The facility is capable of supporting active test article gimbaling during hot fire and is capable of taking side loads of up to 12,000 lbf.

A rolling deck (rated to 8,000 lbm) and crane system (rated to 6,000 lbm) are available to facilitate test article positioning and installation.

3.3.2.2.3.2 Flame Deflector

The facility is equipped with an ablative flame deflector to direct engine plume gases away from the test article and test stand (i.e., toward the east side of the test stand). It is capable of withstanding the pressure and heat loads generated by a 120,000 lbf thrust class engine or combustion device. The deflector is capable of supporting active gimbaling during hot fire. Gimbaling limits will be identified as required based on specific test article thrust levels and plume phenomenology.

3.3.2.2.3.3 Hydraulics System

The facility upgrades will include a hydraulic system for operating facility hydraulically-actuated valves and for supporting test article needs. This system is capable of generating 8 - 15 GPM at 3,000 - 4,500 psig. The system has a 60 gallon hydraulic fluid reservoir with a 5 gallon accumulator.

3.3.2.2.3.4 Deluge System

The facility has a 4,000 GPM, 225 psig MAWP water deluge system for the purpose of limiting damage in the event of a test stand fire and to protect the test stand structure from plume radiant heating during testing. The deluge system covers all four test cell levels, along with the test stand structure, LOX run tank, RP-1 run tank, LOX and RP-1 storage tanks, and the LOX and RP-1 tanker fill headers. All deluge nozzles are remotely operated from the control room.

3.3.2.2.3.5 GN$_2$ Heated Purge System

A GN$_2$ heated purge system is available to facilitate engine drying and RP-1 evaporation. The system is capable of providing 170 °F (maximum) GN$_2$ at 135 psig (nominal) with flow rates of 40 - 60 SCFM (0.05 - 0.07 lbm/sec). Up to four ½ in. test article interface ports are available for simultaneous use.

3.3.2.2.3.6 TEA/TEB Filling System

Facilities are available for filling, storing, and transporting Triethyl-Aluminum (TEA) and Triethyl-Borane (TEB) test article ignition canisters.

3.3.2.2.3.7 Mobile Engine Service Panel

A mobile engine service panel is available to facilitate leak checks on test articles by enabling remote pressurization of the test article using GN_2 or GHe. GN_2 at 100 psig and 800 psig and GHe at 100 psig are available at LOX cleanliness levels (SSC level 1XX). GN_2 and GHe at 100 psig are available at RP-1 cleanliness levels (SSC level 2X). Pressure measurement ports for reading internal engine pressures during pressurization and leak check are included on the panel.

3.3.2.2.3.8   Communications System

The facility communications system has the capability to provide voice communications throughout the E-2 Test Facility.

3.3.3   E-3 Test Facility [47]

The E-3 Test Facility is a versatile test complex available for component development testing of combustion devices, rocket engine components and small/sub-scale complete engines and boosters. The facility currently has two test cells. Cell 1 is a horizontal test stand, which can support horizontal thrust loads up to 60,000 lbf (120,000 lbf impulse load). Cell 2 is primarily for vertical testing with provisions for limited horizontal testing. Cell 2 can support vertical thrust loads up to 25,000 lbf thrust (50,000 lbf impulse load). Cell 2 has a flame bucket below the firing position. The addition of a third test cell (Cell 3) is under consideration.

The facility has the capacity to deliver propellants at low and medium pressures, up to 3,000 psi. All propellant storage, transfer, and run systems for LOX and GOX are cleaned to cleanliness level 1XX per SSC STD 79-001. Similar systems for H_2O_2 are initially cleaned to cleanliness level 1XX and then passivated for H_2O_2 service. The JP systems are initially cleaned to a level 2X and, with exception of the final filter, are maintained at level 3. From the final filter to the test article interface, the run systems are maintained to level 2X. GH_2 is available from the site-wide system and can be delivered with a cleanliness level of 2X.

Single-axis thrust measurement capability is available for both Cell 1 and Cell 2. Currently, 10,000 lbf and 25,000 lbf thrust measurement systems (TMS) are available for use. An additional TMS unit of 60,000 lbf capacity is in the facility upgrade plan.

Test cells 1 and 2 can be occupied at the same time, providing a multiple program capability. Both test cells are adequately illuminated for night time work.

E-3

3.3.3.1 Cell 1 – Horizontal Position

E 3 Cell 1

Cell 1 was primarily designed to test pressure-fed LOX/hydrocarbon fuel, GOX/hydrocarbon fuel, GH_2/GOX, and hybrid rocket motor combustion devices. JP and H_2O_2 run systems will be installed in Cell 1. Cell 1 has two thrust positions. Both positions are capable of supporting horizontal thrust loads of up to 60,000 lbf (120,000 lbf impulse load). The actual thrust capability will depend on the test article mounting position and thrust centerline orientation and can be evaluated on a case-by-case basis. Additionally, Cell 1 has a small component test position capable of supporting 3,000 lbf thrust loads (6,000 lbf impulse load).

Cell 1 is 38 ft. in width by 40 ft. in length and is covered with a roof 25 ft. in height. A 5-ton overhead crane provides lifting capability up to a height of 18 ft. The run tanks are conveniently located to suit the test programs. The table on the next page outlines the existing and planned upgrade commodity supply capabilities for Cell 1. The flow rates listed in the table are nominal. Flow rates higher than those listed in the table can be attained and can be determined on a by-case basis. A 4 in. LOX/LN$_2$ discharge (drain) line is routed east of the cell to a ditch located north of the facility. As part of the facility upgrades, a 4 in. H$_2$O$_2$ vent line will be routed to the north wall for discharge into the containment area. Primary H$_2$O$_2$ liquid discharges will be routed through a catalyst bed for decomposition prior to discharge. Cell 2 has a similar capability (see Section 3.3.3.2).

Cell 1 Commodity Supply Capabilities

Commodity	Pressure (psig)	Temperature (°R / °F)	Flow Rate (lbm/sec)	Supply/ Run Line (in.)	Existing / Upgrade	
LOX	1,500[1]	163 / -297	9[3]	1	☑	☐
LOX	1,500[1]	163 / -297	22	1 ½	☑	☐
LN$_2$	1,500[1]	144 / -316	8[3]	1	☑	☐
LN$_2$	1,500[1]	144 / -316	16	1 ½	☑	☐
GOX	2,200[2]	540 / 80	31	1	☑	☐
H$_2$O$_2$/LOX	3,500	540 / 80 163/ -297	220/112[4]	4	☐	☑
H$_2$O$_2$/LOX	3,500	540 / 80 163/ -297	30/15[4]	1 ½	☐	☑
JP-8	3,500	540 / 80	40	2	☐	☑

1 - Limited by run line components
2 - K-bottle maximum limit
3 - Maximum flow rate for flow-meter
4 - Maximum flow rate limited by velocity (25 ft/sec)

3.3.3.2 Cell 2 – Vertical Position

E 3 Cell 2

Cell 2 was primarily designed to test H_2O_2/JP-8 and rocket motor combustion devices up to 25,000 lbf of vertical thrust (50,000 lbf impulse load). At present, Cell 2 is configured to support testing of 10,000 lbf H_2O_2/JP engines. As part of the planned upgrades, Cell 2 will be capable of testing H_2O_2/JP engines up to 25,000 lbf vertical thrust. The upgrades will also provide the cell with the capability for testing LOX/JP engines. Cell 2 has an additional capacity to test mono-propellant configuration sub-scale combustion devices, such as catalyst beds and components. The following table outlines the commodity supply capabilities for Cell 2. The flow rates listed in the table are nominal. Flow rates higher than those listed in the table can be attained and can be determined on a case-by-case basis.

Cell 2 features a skid based design concept. In this concept, all test specific hardware (run tanks, run lines, and test article) are mounted on a platform that is bolted above the 8 ft wide by 17 ft deep concrete flame bucket. The existing platform contains a 500 gallon oxidizer run tank and a 250 gallon fuel run tank. These are mounted next to each other, south of a 48 in by 48 in flame bucket access hole provided for vertical testing.

Two vertical thrust takeout structures are available, mounted above the flame bucket access hole. One can be outfitted with existing 10,000 lbf thrust single axis TMS, whereas the other is shorter and stiffer, but does not have TMS capability. A vertical thrust takeout structure with a 25,000 lb thrust (50,000 lbf impulse load) rating will be constructed as part of the facility upgrade plan. The maximum vertical thrust rating is limited to 25,000 lbf. Mobile cranes are available to provide lifting capability. Currently, there is no overhead roof on Cell 2, but a roof will be installed as part of the planned facility upgrades.

Cell 2 Commodity Supply Capabilities

Commodity	Pressure[1] (psig)	Temperature (°R / °F)	Flow Rate[2] (lbm/sec)	Run Line (in.)	Existing / Upgrade	
H_2O_2	1,200	540 / 80	49	2	☑	☐
JP-8	1,400	540 / 80	7	1	☑	☐
H_2O_2/LOX	3,500	540 / 80 163 / -297	220 / 112[2]	4	☐	☑
H_2O_2/LOX	3,500	540 / 80 163 / -297	30 / 15[2]	1 ½	☐	☑
JP-8	3,500	540 / 80	40	2	☐	☑

1. Pressure limited by run line components

2. Maximum flow rate limited by velocity (25 lb/sec)

3.3.3.3   Commodity Storage Capacities and Capabilities

This section summarizes the supply and storage capabilities of the E-3 facility. The following table contains a complete listing of E-3 facility pressure vessels.

E3 Facility Pressure Vessels

Vessel Number	Vessel Commodity	Location	MAWP (psig)	Volume (gallon)	Cleanliness Level	Existing/Upgrade	
V-271-LO	LOX	Cell 1	60	600	1XX	☑	
V-272-LO	LOX	Cell 1	2,000	100	1XX	☑	
V-187-JP	JP	Cell 1	3,500	250	2X[2]		☑
V-479- H_2O_2	H_2O_2/LOX[1]	Cell 1	3,500	500	1XX[3]	☑	
V-230- H_2O_2	H_2O_2/LOX[1]	Cells 2 and 3	3,000	500	1XX[3]	☑	
V-455- H_2O_2	H_2O_2/LOX[4]	Cells 2 and 3	4,500	2,000	1XX[3]		☑
V-186-JP	JP	Cells 2 and 3	3,500	250	2X[2]	☑	
V-478-JP	JP	Cells 2 and 3	3,500	1,000	2X[2]	☑	

1.) Vessels are single wall, but can be insulated for cryogenic service.

2.) Vessels are initially cleaned to Level 2X, but are maintained at a Level 3 cleanliness level. Run systems between the final filter and test article interface are maintained at Level 2X.

3.) Vessels are initially cleaned to Level 1XX and are then passivated for H_2O_2 service.

4.) Vacuum jacketed vessel.

3.3.3.3.1   Liquid Oxygen (LOX)

A 600 gallon LOX vessel provides storage capacity for the 100 gallon LOX run tank at Cell 1.

3.3.3.3.2   Gaseous Hydrogen (GH$_2$)

Sitewide GH$_2$ supply is available at E-3 up to 4,000 psig.

3.3.3.3.3   Gaseous Nitrogen (GN$_2$)

Sitewide GN$_2$ supply is available at E-3 up to 4,400 psig, regulated down to 3,000 psig at the test cells. In addition, GN$_2$ can also be supplied by tube bank trailers.

3.3.3.3.4   Gaseous Helium (GHe)

Sitewide GHe supply is available at E-3 up to 4,400 psig. In addition, GHe can also be supplied by tube bank trailers.

3.3.3.3.5   Inert Gas Storage

As discussed in Sections 3.3.3.3.3 and 3.3.3.3.4, storage of inert gases is accomplished by tube bank trailers. High pressure, inert gas storage vessels and associated piping will be installed as part of the facility upgrades. It is planned that the storage system will be capable of supporting a hot fire test of a 25 Klbf, pressure-fed engine with a chamber pressure of 2,000 psia for a 240 second duration.

3.3.3.3.6   Hydrogen Peroxide (H$_2$O$_2$)

H$_2$O$_2$ is currently supplied in either 30/55 gal drums or in bulk from mobile ISO containers. For its storage, an isolated 30 ft by 35 ft area surrounded by a 2 ft high earthen berm is located east of the test cells. A bulk storage capacity for high concentration H$_2$O$_2$ is being installed at the E-3 Test Facility. Additionally, an H$_2$O$_2$ enrichment facility is being added to the storage area. A remote transfer system will be integrated into the E-3 facilities in order to fill run tanks without disconnecting the operating piping system.

3.3.3.3.7   Hydrocarbon Fuel (JP)

JP is currently supplied from a 500 gal portable tank. A 10,000 gallon storage tank and transfer line will be installed by the facility upgrade program.

3.3.3.3.8   Gaseous Oxygen (GOX)

GOX is supplied from K-bottles.

3.3.3.4   Electrical Systems

The capabilities described below are for test cell 1 and 2.   Upgrades to the Data Acquisition and Control Systems (DACS) currently underway and the Test Control Center (TCC) currently in the design phase are anticipated for completion in late Fall, 2001.

3.3.3.4.1   Control System

The E-3 Test Control Center (TCC) located in the E-Complex Test Operations Building (TOB) houses the control system Programmable Logic Controller (PLC), and serves as the central command location for the test conductor and test personnel during test operations.   The control system provides real time control of test article coolant, propellants, and control valves.   It can automatically cycle control valves through a series of predetermined states specified by the customer.   Test article red line and blue line limit monitoring is also performed by the control system prior, during and post test. The customer can monitor the test article's parameters on a computer screen running Wonder Ware Operator Interface (OI) software.   Data from the control system are time tagged with IRIG B and stored for real-time display and posttest analysis.

The new control system can provide the following number of measurement and control channels (each cell) for a test article:

- 50 analog inputs for pressure, RTD, strain, or acceleration measurement
- 64 measurements for thermocouples
- 80 discrete control valves
- 8 variable control valves (4 are set up for hydraulic actuation; 4 are set up for pneumatic actuation)
- 96 discrete indications

Red line monitoring and cut has a response time of 150 – 200 ms.

An upgrade to the existing TCC is underway.   The larger room will provide better operability of the test systems and a dedicated area for customers to observe real-time test data.

3.3.3.4.2   Instrumentation

Facility instrumentation is installed for real time display of facility processes and data recording. Instrumentation provided for the test article will be determined with each customer. The facility also provides the ability to display real-time test article measurements. Each measurement is sampled and stored at 250 samples/second. Instruments that provide test article response information, time critical control functions, or are located at facility to test article interfaces have a typical 5 to 250 ms response depending on the specific instrument.

3.3.3.4.3　High Speed Data Acquisition System

The High Speed DAS (HSDAS) has a high speed 16 bit digitizer and recorder. The digitizer can be configured as follows:

- 62 channels at 50 KSPS (22.5 KHz)
- 32 channels at 100 KSPS (45.5 KHz)
- 16 channels at 200 KSPS (80 KHz)

Typically, the digitizer is configured for 32 channels at 100 KSPS per channel. The data is recorded digitally on Super VHS tape. One channel is reserved for IRIG B for time correction. A second channel is typically used for Time zero (start of test). Event channels are also available.

3.3.3.4.4　Low Speed Data Acquisition System

The LSDAS system has a 128 channel digitizer per cell that share 100 (50 per test cell, but patchable) programmable bridge type signal conditioners. The system provides 16 bits of resolution with a throughput of 200 KSPS. These systems provide reconfiguration flexibility and hardware/software common with other existing test systems.

3.3.3.4.5　Data Processing

Data processing is provided for the LSDAS and HSDAS. The LSDAS data is converted to Engineering units and processed into the E Complex WinPlot file format. WinPlot is a NASA-developed program for plotting measurement and calculated values. A quicklook program is also available for on-line analysis. Data backup is provided on DAT or CD-ROM formats.

The HSDAS data is available in both time domain and frequency domain formats. Digital filtering algorithms are available for processing. Data is provided on CD ROM for processed data and Super VHS for raw dynamic data.

Provisions can be made for specialty instrumentation such as Optical Deflectometers and Speed Sensors.
Posttest, off -site data transfer is available for both systems through the Internet. Data security is provided through password protection. Data encryption is also available.

3.3.3.4.6　Power

AC - The facility power system provides single and three-phase power at 480VAC, 277VAC, 208VAC, 220VAC, and 120VAC. Uninterruptible power supply (UPS) systems provide a minimum of ten minutes of 120VAC power to test critical systems.

DC - 28V DC is provided for control system I/O.

3.3.3.4.7   Video Systems

E-3 utilizes four low speed video cameras for normal testing. Each camera has pan, zoom, focus, iris, and tilt capabilities. The camera enclosures, boxes, and conduits are purged with GN to maintain compliance with the appropriate electrical hazardous classification requirements. Cameras 1, 3, and 4 are permanently mounted to the canopy columns at the Northwest, Northeast, and Southwest columns respectively. Camera 2 is portable. There is one portable high speed camera available at the test stand. It is capable of recording up to 500 frames per second.

3.3.3.5   Ancillary Facility Systems

3.3.3.5.1   Deluge System

E-3 uses a 6 in. potable water system for the purpose of providing deluge cooling water and limiting damage in the event of a fire in a test cell or propellant storage/handling area. The system includes spray nozzles installed on the roof of Cell 1 and two water cannons situated between Cell 1 and Cell 2. The E-3 deluge system covers all existing test cells, oxidizer vessels, fuel vessels, and all tanker fill headers. The supply system will be integrated with the E-Complex deluge water supply system as part of the facility upgrades. All areas can be remotely operated from the control room. In addition, the two water cannons can be locally operated.

A DI water system is installed for flushing H_2O_2 run tank and run lines.

3.3.3.5.2   Plume Impingement Area

Plume impingement areas are provided for both Cell 1 and Cell 2. The areas feature thrust deflectors that are made of refractory concrete ablative material for the purpose of minimizing the effects of heat and acoustic loads generated by a combustion device plume. The material has excellent thermal properties for high temperature applications.

In Cell 1, the plume impingement area extends northward from the test stand. The plume is deflected away from the concrete foundation and metal grate that form the floor of this test cell. High heat-load test articles are positioned such that the nozzle exit area for plume is outside the main cell structures.

In Cell 2, the plume is directed first downward into the flame bucket and then redirected northward. For this purpose, the plume impingement area at the bottom of the flame bucket is sloped.

3.3.3.5.3   Hydraulics

A 3,000 psig hydraulic system for actuating facility and special test equipment valves will be installed under the facility upgrade program.

3.3.3.5.4   Communications

The E-3 facility is equipped with headset communication boxes in numerous locations to allow voice communication between the TCC and the test facility during operations. The facility is also outfitted with an intercom system.

3.3.3.5.5 H_2O_2 Vapor Monitoring

A remote H_2O_2 Vapor Monitoring System will be integrated into the E-3 test area as part of the facility upgrades.

3.3.3.5.6 H_2O_2 Containment

Per the facility upgrade plan, a containment pond for H_2O_2 discharge and dilution will be built and integrated into the E-3 Test Facility to support test programs and commodity bulk storage.

3.3.3.5.7 H_2O_2 Enrichment Facility

The Hydrogen Peroxide Enrichment Skid is a transportable processing plant that enriches aerospace grade H_2O_2 from 90% to 98% final concentration. The Skid is a self contained unit that houses all of the tanks, lines, coolants, crystallizers, instrumentation and program logic controllers to perform the concentration operation nearly autonomously. Located near E-3, it will be operational in the late 2001/early 2002 timeframe. The unit is positioned on a concrete slab but it is possible for it to be relocated for remote use.

3.3.4 E-4 Test Facility

The newest addition to the E-Complex will be the E-4 Test Facility, scheduled to be complete by December 2003. The E-4 Test Facility will be capable of providing a low-pressure supply of JP-7 and LOX to test articles having a thrust in the horizontal plane up to 50,000 lbf (maximum). The facility's design is especially suited for the testing of Rocket Based Combined Cycle (RBCC) test articles.

The E-4 Test Facility design also allows for the growth of test capabilities to meet future testing requirements of RBCC and potentially other engine concepts. The future development of the E-4 Test Facility is envisioned to occur as follows:

1) The incorporation of an LH_2 and an H_2O_2 propellant supply capability to support the testing of power packs and engine systems up to 50,000 lbf thrust (maximum).

2) The addition of a Ram Air test capability up to Mach number 0.8 for engine systems having a maximum thrust of 50,000 lbf.

3) Upgrading the necessary propellant and structural entities to support the testing of power packs and engine systems up to 500,000 lbf thrust (maximum).

4.0 PROPULSION TEST SUPPORT

SSC maintains a number of facilities and provides specialized services required for the direct support and operation of test facilities. Included are the Cryogenic Propellant Storage Facility, High Pressure Gas Facility (HPGF), High Pressure Industrial Water (HPIW) Facility, emergency power-generation facilities, and electrical distribution systems. Additional information related to propulsion test support facilities and services is provided in the following sections.

4.1 Cryogenics Operations [27,43]

Cryogenics Operations provide LOX and LH_2 for the engine testing programs at SSC that require liquid propellants. This support includes operating and maintaining an LH_2 Storage and Transfer Facility, a LOX Transfer Facility, three LH_2, and six LOX barges. The LH_2 barges can be transported by tugboat to a supplier facility to be filled with propellants, but this has not been done since 1997. The LH_2 barges can also be filled at the LH_2 Storage and Transfer Facility from the storage sphere or directly from tank trailers through six off loading stations. The six LOX barges are filled at the LOX Transfer Facility directly from the tank trailers through the six off loading stations. The Cryogenics Operations technicians provide standby and real time support to ensure all LOX and LH_2 facilities, barges, and equipment are ready and function properly for engine testing. They also perform propellant sampling, vessel certification, calibration, preventive maintenance, vacuum maintenance, and corrective maintenance.

4.1.1 Liquid Oxygen Service

LOX services at the Cryogenics Operations area include transfer operations from vendor delivery trailers to SSC storage barges, barge-to-barge transfer, and land-base storage (downmoded). Additional details related to LOX transfer operations and storage barges are provided in the following sections.

Liquid Oxygen Storage and Transfer Facility

4.1.1.1   Transfer Operations

Delivery of LOX to the Cryogenics Operations area storage and transfer facility at the cryogenic propellant dock is supplied by vendor tank trucks. There are two stations, each with three trailer positions, at the dock for downloading LOX from vendor trucks. The trucks each contain approximately 4,400 gal. and are downloaded via 3 in. flex hoses, which are connected to a single 8 in. pipe that connects to each LOX barge position. Downloading time per trailer averages 1 hour (hr) and 30 min, with simultaneous downloading of up to six trailers. Each position can be isolated by a series of valves that allow each barge to be singularly filled from the vendor trucks at 20 - 30 psig per truck, or to transfer LOX from one barge to another during topping or downloading operations. A vaporizer on each vendor tank truck is used to pressurize the trailer storage tank to 20 - 30 psig and thus effect LOX transfer to a LOX barge.

Provision for storage and transfer of LOX between a 460,000-gal dockside storage tank and transfer vehicles (e.g., barges and trucks) was discontinued. LOX losses incurred during storage and repeated transfers were excessive and cost prohibitive. Hence, LOX is presently transferred directly from the vendor tank trucks to the LOX barges through the propellant dock transfer piping.
Downmoded equipment includes a 460,000 gal storage tank and a vaporizer system capable of pressurizing the LOX storage to 20 psig for transfer to the LOX barges.

4.1.1.2   Storage Barges (52,53)

Six LOX transfer barges are used to transport LOX from the Cryogenic Operations area to the Test Complexes. Dock facilities for two barges exist at the A-1 and A-2 test stands, and docking facilities for three barges exist at the B-1/B-2 test stand. Two additional barge positions are available at the "B" Test Complex, but have no piping or mooring devices.

Each barge has a total storage capacity of 105,000 gal gross of LOX with 95,000 gal usable at 40 psig. Barge-mounted pumps transfer LOX from the barges through the dock transfer system into the test stand storage/run tanks. The barges are also used to receive LOX from the storage/run tanks in the event they have to be emptied. Below is a list of the barge's capabilities.

    a. Two transfer pumps rated at 1,250 gal/min, each at 250-350 psig, nominal capacity at 3,600 revolutions per minute (rpm) for a total pumping capacity of 2,500 gal/min.

    b. One topping pump, rated at 250 gal/min at 250-300 psig at 3,600 rpm. (Topping pump system downmoded and removed from some barges)

    c. Onboard GN_2 storage sufficient to provide for necessary valve operations and purging.

d. A deluge system supplying industrial water for fire protection. Water for this system is available when the barge is docked at the storage facility or the test stands.

e. Barge-to-dock connections for LOX transfer including:

   (1)   Main transfer line

   (2)   Topping transfer line (where applicable)

   (3)   Control and instrumentation

   (4)   Main pump's electrical power

   (5)   Topping pump's electrical power

   (6) GN_2 supply at 2,250 psig dockside, reduced to 750 psig on barge (for control-valve operations)

   (7)   Barge electric power system

   (8)   Deluge water system.

4.1.2 Liquid Hydrogen Service

LH_2 services at the Cryogenics Operations area include barge-to-barge LH_2 transfer, transfer operations from vendor delivery trailers to SSC storage barges, and barge docking. Additional details related to LH_2 transfer operations and storage barges are provided in the following sections.

4.1.2.1   Transfer Operations

Provisions for delivery of LH_2 by barges or trailer trucks are available at the LH_2 storage and transfer facility in the Cryogenics Operations area. The facility consists of six trailer transfer stations, a 600,000 gallon storage sphere, and three barge docks. Two trailer transfer stations can be modified to accept rail cars. Two barge docks are complete transfer facilities, while one requires a facility interface for the liquid transfer line. Six vendor trailers can off-load simultaneously at 40-60 psig via 2 in. flexhoses connected to a 4 in. header. The 4 in. header connects to an 8 in. sphere/barge transfer line. Off-loading of vendor trailers averages one hr and 10 min with each trailer containing approximately 12,000 gallons of liquid. Total connect, off-load, and disconnect takes approximately two hrs and 30 min. Transfer rates per trailer is approximately 200 gpm. A vaporizer on each vendor trailer is used to pressurize the trailer storage tank to 40-60 psig and thus affect LH_2 transfer to the storage sphere and/or barge.

Barges can be filled directly from trailers and/or the storage sphere. Pressure transfer from the sphere to barge is at 9 psig with a flow rate of approximately 3,000 gpm. Barge loading takes approximately one hr and 30 min. Total connect, load, sample, and disconnect takes about nine hours. Barges can also be off-loaded into the sphere in case of emergency. Barge to barge transfer is also available.

4.1.2.2 Storage Barges (54)

Three LH_2 transfer barges are used to transport LH_2 from the LH storage and transfer facility to storage/run tanks at the A-1, A-2, and B-1/B-2 test stands. Each barge has a total capacity of 270,000 gal and a usable capacity of 240,000 gal at 70 psig. A barge-mounted LH_2 vaporizer provides ullage pressure (60-70 psig) to transfer LH_2 from the barge through the test stand dock-transfer facilities to the storage/run tank at a nominal flow rate of 5,000 gal/min. The barge can also be used to receive LH_2 from the storage/run tank at an approximate rate of 3,500 gal/min in the event the tank has to be emptied. Each barge is equipped with the following:

a. One LH_2 vaporizer rated at 92.3 psig design pressure and 4,320 pounds per hour (lb/hr) at 70 psig.

b. Onboard GN_2 storage sufficient to provide for necessary valve operations and purging during transit on-site (5 hr transit to and from the vendor's facility, as required).

c. One deluge system requiring 4,400 gal/min of industrial water at 80 psig for fire protection. Deluge system hookup is available at each test stand docking area and at the LH_2 storage and transfer facility.

d. Barge-to-dock connections, which interface with the following dockside systems during LH_2 transfer:

    (1) Main transfer line

    (2) GH_2 gas vent line

    (3) GN_2 purge system at 2,250 psig dockside, reduced to 750 psig on barge

    (4) Deluge water system

    (5) Control and instrumentation

    (6) Electrical power supply

    (7) Electrical grounding system

4.2 High Pressure Gas Facility [4,24,27,37,38,42,43](101)

The High Pressure Gas Facility (HPGF) receives, stores, and distributes the HP gases required to support engine testing programs and other SSC-assigned missions. The HPGF supplies air, GHe, GN_2, and GH_2 to the test stands.

The HPGF support is scheduled on a two-shift basis providing operations, preventive and corrective maintenance, and periodic sampling, certification, and calibration (systems, subsystems, and components). Additional information relating to the HPGF is contained in the sections that follow.

4.2.1 Air Supply System (46)

The High Pressure Air (HPA) system is designed to produce missile-grade air and meet SSC-79-002. System pressure is maintained between a minimum of 1,500 psig and a maximum pressure of 2,800 psig; MDWP is 3,500 psig.

Atmospheric air is supplied by individual, oil bath filters at the roof level of the HPGF. The atmospheric air is compressed by one five-stage compressor, two identical Clark compressors, and three Cooper compressors. Each compressor is rated for 850 SCFM at 3,500 psig, powered by 500 horsepower (hp) electric motors.

Cooper Compressor at HPGF

Two air-drying filtering systems are interconnected to allow any compressor to discharge to either system. The absorption unit consists of beaded activated charcoal to remove hydrocarbons and compressor oil. The air dryer (Letrodryer) consists of two HP containers filled with desiccant beads to remove moisture.

The air is filtered through a bank of wafer, felt-type, 20 μ filters and a set of 20 μ sintered backup filters to remove accumulated particles. The missile-grade air is then distributed through a 3 in. diameter transfer line to the site-wide systems. The system pressure is maintained by operator observation of the system pressure and the starting and stopping of the compressors. HPA is stored at individual site locations and maintained at pressure via cross-country gas supply from the HPGF.

4.2.2 Helium Supply System [43](40,41)

The HP He system is designed to produce HP He with a dewpoint of -100 °F minimum, and a maximum hydrocarbon content of 10 ppm by weight. System pressure is maintained between a minimum 2,000 psig and a maximum 4,500 psig with an MDWP of 4,500 psig.

GHe is delivered to SSC in tube bank trailers and downloaded into two LP storage vessels at the HPGF. The table provides storage capacity and working pressure for each vessel.

Vessel Locator Number	MDWP (psig)	Certified Pressure (psig)	Water Volume (ft^3)	Storage at 450 psig, 70 ° F (scf)
V-20-HE	520	450	10,000	350,000
V-21-HE	520	450	10,000	350,000

HP GHe is provided to the site-wide systems by compression of GHe from the storage vessels. Three identical Clark, oil-lubricated, reciprocating compressors, each rated for 200 ft^3/min at 6,000 psig, driven by 125 hp electric motors, discharge to a common header. Two identical Henderson compressors, each rated for 250 – 400 ft^3/min at 6,000 psig, also discharge to the common header. The GHe is passed through a separator and an adsorber to remove oil contaminants. The GHe is then filtered through a 20-µ filter and three, parallel, molecular sieve, cartridge-type purifiers that remove water vapor. The GHe is then distributed to the site-wide systems. Excess trailer-supply GHe is transferred directly into the HP GHe system, using the same process as the transfer from LP storage. The GHe is then distributed through a 1 1/2 inch diameter transfer line to the site-wide systems.

Three Clark Compressors at HPGF

The system pressure is maintained via operator observation by manually starting and stopping the compressors. All HP GHe is stored at individual site locations and maintained at pressure via the cross-country gas supply from the HPGF.

4.2.3 Nitrogen Supply System [43](28-31)

The HP GN_2 system is designed to produce HP GN_2 with system pressure maintained between a minimum of 2,400 psig and a maximum of 4,400 psig with an MDWP at 6,000 psig. LN_2 is delivered to SSC by vendor tank trucks and downloaded through 40 μ filters into two LP storage vessels at the HPGF. The LN_2 transfer is performed by vendor tank truck self-pressurization systems or with facility pumps [two pumps, each rated for 200 gal/min at 25-30 psig (P1LN and P2LN), driven by 5 hp electric motors]. The following table provides storage capacity and working pressure for each vessel.

Vessel Locator Number	MDWP (psig)	Certified Pressure (psig)	Water Volume (gal)
V-112-LN	33 (ullage)	33 (ullage)	63,250
V-173-LN	25 (ullage)	25 (ullage)	27,000
V-174-LN	50 (ullage)	* (ullage)	15,000

* Downmoded (not in service)

The LP storage vessels are maintained at a standby pressure of 2-3 psig, with a maximum working pressure of 22-25 psig during operations. LN_2 (from the storage vessels) is pumped with three centrifugal booster pumps (each rated for 60 gal/min) at 65-75 psig, which are driven by 7.5 hp electric motors, through a 40 μ filter, to the suction of six Kobe positive-displacement pumps (each rated for 15.6 gal/min at 6,000 psig), which are driven by 60 hp electric motors. The Kobe pumps discharge through hot-water vaporizers [each rated for 1,450 standard cubic feet per minute (scfm) at 6,000 psig] to gasify the LN_2. The GN_2 is then distributed through a 4-in.-diameter transfer line to the site-wide systems.

The system pressure is maintained via operator observation by starting and stopping the pump systems. All HP GN_2 is stored at individual site locations and maintained at pressure via the cross-country gas supply from the HPGF.

Kobe Pumps at HPGF

4.2.4 Hydrogen Supply System (7-9)

The HP GH_2 system is designed to produce HP GH_2 with the system pressure maintained between a minimum of 2,200 psig and a maximum of 3,000 psig with an MDWP of 6,000 psig, currently limited to 4,000 psig by LH_2 pumps.

LH_2 is delivered to SSC by vendor tank trucks and downloaded into the HPGF LP storage vessel. The LH_2 transfer is performed by the vendor tank truck self-pressurization systems or with onboard tank truck pumps. The storage capacity and maximum allowable working pressure for the HPGF LP vessel is in the following table.

Vessel Locator Number	MDWP (psig)	Certified Pressure (psig)	Water Volume (gal)
V-267-LH	150 (ullage)	150	20,000

The LP storage vessel is maintained at a maximum working pressure of 95-100 psig. LH_2 from the storage vessel is pumped with two, cryogenic, VJ, reciprocating pumps (each rated for 5.2 gal/min at 4,000 psig), which are driven by 30 hp electric motors through aluminum-finned, ambient-air vaporizers (each rated for 35,000 ft³/hr at 4,000 psig). The GH_2 is then distributed through a 2 in. diameter transfer line to the site-wide systems.

The system pressure is maintained automatically by a PLC-based control system operating the pumping equipment. All HP GH$_2$ is stored at individual site locations and maintained at pressure via the cross-country gas supply from the HPGF.

Hydrogen Storage Tank at HPGF

4.2.5 Auxiliary Tube Bank Storage

There are seventeen movable, trailer-mounted tube bank storage vessels that are used at locations having no permanent HP gas supply. The tube bank trailers can be used to store HPA, GHe, or GN$_2$. The tube bank trailer numbers and design capacities are given in the chart on the next page.

Tube Bank Trailers at HPGF

Trailer Number	Gas Service	Design Capacity (scf of air)

71

91-1	GHe	78,000 scf at 6,000 psig
91-2	GN$_2$	78,000 scf at 6,000 psig
91-3	GHe	78,000 scf at 6,000 psig
91-4	GH$_2$	78,000 scf at 6,000 psig
91-7	GN$_2$	50,000 scf at 2,400 psig
1589	GH$_2$	46,200 scf at 3,600 psig
1597	GHe	48,400 scf at 3,600 psig
2542	GN$_2$	39,200 scf at 2,400 psig
6748	Air	34,230 scf at 2,400 psig
6749	GN$_2$	33,600 scf at 2,400 psig
6750	GN$_2$	33,600 scf at 2,400 psig
8205	GHe	44,000 scf at 2,400 psig
8343	GHe	31,920 scf at 2,400 psig
9096	GN$_2$	57,600 scf at 2,400 psig
9403	GN$_2$	46,400 scf at 2,400 psig
9404	GHe	54,720 scf at 2,400 psig
9960	GHe	37,240 scf at 2,400 psig

4.2.6 Cooling-Water System

Cooling water is utilized at the HPGF for the air-compressor, air-dryer unit, and for He-compressor cooling. Water is circulated in a closed-loop system through equipment, heat exchangers and then cooled in separate Marley cooling towers.

The three induced-draft cooling towers are each equipped with a propeller fan, driven by a 10 hp electric motor. Water is pumped from each cooling tower to the required equipment items by three electric motor-driven pumps. Two are 10 hp and one is 15 hp. Each pump is capable of 300 gal/min at a 200 ft. discharge head. The pumps are continually providing water circulation. Cooling tower fans are activated when the water temperature exceeds 80 °F and are deactivated when the water has cooled to 70 °F.

Cooling Water Towers at HPGF

4.3   High Pressure Industrial Water Facility [5,7]

The HPIW facility furnishes water to the "A" and "B" Test Complexes for test stand deflector coolant, fire protection (deluge), and diffuser operation (A-2 and B-1). It also furnishes water for fire protection of the propellant barges (LH$_2$) at the test stands. The HPIW facility also houses electrical power-generation equipment for emergency backup supply to the test complexes. Additional information relating to the HPIW facility is contained in the following sections.

4.3.1 Industrial Water System [7]

Four pumps, with a capacity of 5,000 gal/min per pump, supply an 800 ft. diameter, 66 million gallon reservoir (26 M-gal usable), via a 42 in. diameter transfer line from the main access canal. The reservoir can also be supplied from three deep wells, two of which are currently downmoded.

The HPIW pumping system consists of two electric motor-driven pumps and 10 diesel motor-driven pumps. One electric motor-driven pump serves as a jockey pump to maintain system pressure during non-supply operations. The jockey pump is driven by a 50 hp motor and provides 200 gal/min at a 495 ft. discharge head. The other electric motor-driven pump is driven by a 500 hp motor and provides 3,000 gal/min at a 495 ft. discharge head to supply small usages (e.g., test stand run tank fire protection, barge deluge). The 10 diesel motor-driven pumps are single-stage, double suction, centrifugal pumps, with individual ratings of 33,385 gal/min at a 495 ft. discharge head (225 psig) when operating at 905 rpm. Diesel fuel storage consists of a two 25,000 gal double-wall tanks, which are sufficient to provide a 20 hr. supply with all engines operating at their rated load.

73

SSC High Pressure Water Facility

Supply from eight of the HPIW pumps is discharged into an 84 in. diameter manifold, and pumps 9 and 10 discharge into a 112 in. diameter manifold, which supplies a 75 in. diameter transfer line (flow rate: 212,000 gal/min) to the "A" Test Complex and a 96 in. diameter transfer line (flow rate: 290,000 gal/min) to the "B" Test Complex. The 75 in. diameter transfer line to the "A" Test Complex is branched to two 66 in. diameter lines, one each to the A-1 and A-2 test stands. Four additional stub-outs exist in the 112 in. diameter manifold.

At T minus 1 hr. of a test countdown, the pumps are powered-up for the test stands as follows: seven pumps for A-1, eight pumps for A-2, and nine pumps for B-1 and B-2.

1 of 10 HPIW Pumps

10 Nordberg Diesel Engines That Drive The Pumps Pictured Above

4.3.2 Emergency Power-Generating System [7]

The HPIW emergency power-generating system is capable of providing emergency electrical power to the Test Complex (A-1, A-2, and B-1/B-2 test stands, TCCs, and the DAF). There are four, 1875 kVA [1500 kilowatt (kW)], diesel-driven generators located at the HPIW Facility. These four generators can be synchronized with, or operated independently of, the utility-fed circuits 11 and 21 feeders that source only the Test Complex. The emergency generators are used when the frequency or stability of utility power is threatened by inclement weather. The Test Conductor or Instrumentation personnel request generator support. These generators are rated to support the Test Complex until all post-test operations are complete.

Four Diesel-Driven Generators at HPIW

4.4   Site Electrical Distribution [30](51)

SSC's electrical service is supplied by two 115 kV utility lines feeding the double-ended, 13.8 kV main substation at the south side of the site. Each 115 kV line feeds a 35 MVA transformer. The secondaries of the two 35 MVA transformers are designed to be normally separated, but can be tied together with a tie circuit breaker. The two secondaries each feed a group of 13.8 kV feeder circuit breakers (assigned in dual pairs).

SSC A and B Complex distribution circuits are run as dual circuits, with an automatic or manual transfer system at the load ends, which permits rapid restoration of service if one of the feeders should fail. Each of the 13.8 kV, feeder circuit breakers' trip circuits is set at approximately 400 amperes, permitting a maximum load of about 9500 kVA from any 13.8 kV feeder.

The dual-feed system will support the entire connected electrical load from one feeder, if required. See Table 4.4-1 for transformer ratings.

Identifier*	Capacity Out (kVA)	Voltage In (kV)	Voltage Out (V)
1	35,000	115	13,800
2	Removed	Removed	Removed
3	1500	13.8	480
4	1000	13.8	480
5	750	13.8	480
6	500	13.8	480
7	300	13.8	480
8	30	13.8	208
9	3750	13.8	4,160
10	500	13.8	2,400
11	500	13.8	480
12	2000	13.8	480

Table 4.4-1   Transformer Ratings

5.0 TEST AND TECHNICAL SERVICES

The Test and Technical Services Contractor (TTSC) provides on-site support to NASA and the resident agencies at SSC. Areas of support relating to propulsion testing include Test, Engineering & Operations (TE&O), Systems Engineering & Advanced Technology (SE&AT), Science Laboratories, and Information Systems.

5.1 Test, Engineering & Operations (TE&O)

TE&O provides engineering and technical services for the SSC test complexes and the test support facilities. These services consist of Test Stand Operations, Test Engineering, and Test Operations Support.

5.1.1 Test Stand Operations

Test Stand Operations consist of engineering and technical test teams, which provide engineering and technical services to process and install test articles, conduct tests, and remove test articles. Operations teams configure test facilities, control systems, instrumentation, and data acquisition systems for test firings. They also perform post-test data processing, data analyses, and data validation for data package delivery to the test program customers.

5.1.2 Test Engineering

Test Engineering provides mechanical and electrical engineering services to Operations at the test complexes and supporting facilities that are used for preparing and testing liquid, solid, and hybrid propulsion systems, components, and special equipment at SSC. These services involve: establishing and supporting preventive and corrective maintenance activities; design support for refurbishment and/or modification of technical facilities on local and major Construction of Facilities (C of F) projects; support during fabrication, installation and activation of test facilities; support on anomaly investigations and safety hazard analyses. Test Engineering also provides mechanical and electrical engineering services on-demand to other government agencies, contractors, and tenants at SSC.

5.1.3 Test Operations Support

Test Operations Support consists of mechanical and electrical technical support personnel that provide corrective and preventative maintenance support on site-wide technical systems and test complex specific systems. Test Operations Support provides technical services to implement local, major C of F, and commercial "demand" type refurbishment/modification projects that make programmatic and facility upgrade changes to test and technical systems. This section fabricates, installs and activates the test facility equipment; performs anomaly investigations and supports safety hazard analyses. Test Operations Support also provides technical services on demand to other government agencies, contractors, and tenants at SSC.

5.1.3.1   Test Operations Electrical Support

Test Operations Electrical Support provides services for the following electrical test facility systems at SSC:

a. Test Warning Lights System
b. Oral Warning & Public Address System
c. Intercom Systems
d. Cryogenic Systems (LOX & LH_2 Barges/Dock electrical systems)
f. Instrumentation (electronic sensors, measuring equipment)
g. Special Lighting Systems (test warning stations, video camera)
h. Electrical Power Conditioning (GSE inverters, battery chargers, and batteries)
j. Special Power Systems (test article & STE AC/DC power)
k. Electrical Power Generator (Test Complex Backup)
l. Hazardous Gas Leak & Fire Detection Systems (Hydrogen; etc.)
m. Process Control Systems (PLCs, Data Logging, Monitoring)
n. Supervisory Control & Data Acquisition Systems (SCADA) (HPGF automation systems for gas compressors and vaporizers)
o. Flow Measurement Systems (FMSs) (automatic logging and reporting of flow and & volume usage of test complexes inert gasses)
p. Data Acquisition & Control Systems (DACS) (PLCs, signal-conditioners, high speed and low speed multiplexers, tests console displays)
q. Video Systems (High Speed & Low Speed Video cameras and recorders)

During test firings, additional electrical support personnel are provided to support to the Operations team to correct electronic controls, instrumentation, and electrical malfunctions as they occur during the test countdown to minimize test delay and test scrubs.

5.1.3.2   Test Operations Mechanical Support

Test Operations Mechanical Support provides services for the following mechanical test facility systems at SSC:

a. Tubing Fabrication Equipment & Tools (Bending and end-finishing capability up to 1 ½ inch tubing)
b. Pipe Threaders (Portable equipment with capability up to 4 in.)
c. Hydraulic-Jacks (Lifting capabilities up to 50 tons)
d. Assorted Mechanical Tools (Spreaders, cable cutters, wrenches)
e. Portable electrical generators, heaters, air compressors, hydraulic pumps
f. Hoisting Equipment (Full array of proof-tested/certified rigging equipment)
g. Cryogenic Insulation Installation (Piping and components)
h. Diesel Engine Operators (Diesel engines up to 4,600 BHP)
i. Panel Fabrication (Sheet metal products and tooling)
j. Vacuum Systems (Pumps, vacuum-jacketed piping, monitor gauges)
k. Hydraulic Systems (Pumps, flex-hoses, tubing)
l. Cryogenic Systems (Installation experience on cryogenic valves, piping, and equipment)

5.2 Systems Engineering & Advanced Technology

Systems Engineering & Advanced Technology (SE&AT) provides engineering and CAD drafting services in the functional areas of Systems Engineering and Design Analysis, Advanced Technology, and Program Management Support. These services are associated with a wide variety of test articles including solid, hybrid, and liquid propellant type propulsion engines and systems, as well as individual engine components, special equipment, and/or materials tests performed at SSC.

5.2.1 Systems Engineering and Design

Systems Engineering and Design Analysis (SE&DA) provides electrical and mechanical systems engineering, design, and analytical services in direct support of testing activities. These services include: preliminary test requirements analysis; test facility capability improvement efforts and cost estimates; test article to test stand interface analysis; test operations support for independent test analysis and evaluation.

SE&DA services provided during the preliminary stages of testing activity include: design analysis and review; system trade studies (e.g., cost, schedule, safety); systems analysis and modeling, including performance and requirements analysis; technology risk assessments.

Test facility preparation support provided by SE&DA includes design review, design analysis, dynamic fluid flow simulation, propellant transfer systems, piping flexibility, software systems requirements analyses, environmental conditioning systems, material compatibility analysis, thrust system design and analysis, and structural analysis and control.

During test operations, SE&DA provides analytical services to evaluate changes to test requirements and to resolve real-time test system anomalies. Post-test services include independent test analysis and evaluation of test data to support the overall test verification process.

SE&DA provides design-engineering services for the analysis, design, fabrication, installation, and operation of technical facilities and equipment used for testing. Specific design activities include structural/stress analysis (including FEM/A), fluid flow analysis instrument calibration for hydrological applications, and hydrology experimentation and sensor development.

Computer Aided Design and Engineering (CAD/E) methods, including AutoCad, ClarisCAD, Pro/Engineer and Pro/Piping, are used for design and documentation. Engineering analyses are performed using computer techniques including finite element model simulation and nonlinear analysis methods. Fluids analyses are made using either the SUPERFLO or the EASY5X with ROCETS programs. Heat transfer and stress analysis are made using ANSY, Algor, COSMOS, NASTRAN, SCINDA and Pro/MECHANICA programs. Data acquisition and analysis are performed using

LabView. The program EXTEND is used for process simulations. Several other programs are available and are used where they facilitate the task.

5.2.2 Advanced Technology

Expertise exists within this section for electro-optical instrumentation and test measurement system design & integration, data reduction and analysis, and motor specific constituents data base development. Leading edge technology has been developed that can identify specific engine component problems via non-intrusive measurement of plume constituents. Expertise is also available in the areas of plume modeling, acoustic prediction, CFD analysis, thermal analysis, hydrogen leak detection, and hydrogen fire detection and imaging. Support is also provided for thermal imagery applications.

5.2.3 Program Management Support

The Program Management Support Section provides NASA with full program support functions including Configuration Management (CM), Lead Center Development, Project Engineering, and Stennis New Business Development. This section supplies NASA with a single point of contact for program support.

Configuration Management is provided for NASA Propulsion Test technical systems. The CM functions include Configuration Control Board Secretary, Change Request development and processing, document control, and general support to the Board Chairperson.

The Project Engineering function is also provided through this office. Project engineering furnishes NASA with a single point of contact for project integration and coordination of technical support to test project activities. This function is responsible for status reporting to LMSO management and NASA project management on SSC projects.

This section provides support to the Stennis New Business Development Office, which is tasked with identifying and capturing new propulsion testing opportunities for SSC. This section also develops and maintains technical literature to promote SSC's propulsion testing capabilities.

This section also provides support to the NASA Lead Center group. This support includes Rocket Propulsion Test Management Team (RPTMB) and National Rocket Propulsion Test Alliance (NRPTA) activities.

In addition to the services listed above, the Program Management Support section is structured to provide NASA with any other necessary program support to meet the testing customer's needs.

5.3   Laboratory Services

Laboratory services are provided to NASA and other agencies as well as commercial organizations through the Science Laboratories, Environmental Services Laboratory, Measurement Standards & Calibration Laboratories, and Prototype Development Laboratory.

5.3.1 Science Laboratory Services

5.3.1.1   Environmental Laboratory

The Science Laboratory Services Environmental Laboratory consists of a fully equipped analytical laboratory, and facilities for research and development. Performance of the laboratory plays a critical role in determining the outcome of an investigation in terms of the accurate and precise characterization of samples submitted for analysis. Substandard performance by a laboratory can result in untimely delays, unnecessary expenses, incorrect remedial action, and/or fines as well as damage to the site's reputation with its neighbors. The Environmental Laboratory requested and received EPA Certification in 1996. The certification consists of an annual performance-based study. Each year, the results of the study are reported to the state's certifying officer who then issues a certification status report. Currently, the laboratory is certified to perform 85 of the 89 certifiable analyses.

As a convenience to laboratory customers, a variety of reporting capabilities have been developed. The Environmental Laboratory produces two levels of reporting (1) Contract Laboratory Program (CLP) reports, and (2) customized level II reports which are tailored specifically to meet individual customer needs. CLP reporting follows very rigorous EPA guidelines for data quality assurance and control, which sets this level of reporting apart from the typical level II report. Most environmental laboratories do not perform to the extreme level of quality control/assurance measures required of CLP reporting.

Additional information about the SSC environmental program can be obtained from the environmental website www.ssc.nasa.gov/environmental or Stennis Procedures and Guidelines 8500.2, Environmental Operations and Implementation Program.

5.3.1.1.1 Environmental Capabilities

The laboratory has comprehensive experience in environmental services including:

- Chemical analysis of all environmental matrices (air, water, soils, etc.)
- Hazardous and industrial waste characterization
- Preparation and implementation of environmental sampling plans
- Lead/paint remediation analysis/study
- Oil and sludge analysis
- Flora and fauna study
- Wetland mitigation management

Laboratory expertise is based on in-depth knowledge of federal and state regulations and the permitting and implementing requirements established for SSC. All aspects of environmental management, from permitting assistance to sampling, analysis, and statistical evaluation are supported.

5.3.1.1.2 Analytical Services

The Environmental Laboratory has been responsible for the sampling and analysis of the wastewater and drinking water at SSC since its inception. The responsibility has since been expanded to include sampling and analysis of the air, groundwater, soil, and vegetation. Sample collection procedures, preparation, and analysis follow published EPA procedures and are regulated under extensive quality assurance and control programs. Statistical evaluation of the analytical data is used in meeting permitting requirements, preparation of regulatory reports, and research interpretation. The following table consists of major equipment items and the capabilities gained from them.

INSTRUMENT	FUNCTION
Gas Chromatograph (3)	Used to determine part-per-billion (ppb) level concentrations of herbicides, pesticides, polychlorinated BI-phenols (PCBs), and total petroleum hydrocarbon (TPH)
Gas Chromatograph/Mass Spectrometer (2)	Allows for the analysis of soil, water and air for volatiles and semi-volatile compounds
Atomic Absorption Spectrometer	Allows for the analysis of most environmental matrices for metals and cations
Inductively Coupled Plasma Spectrometer (2)	Analysis of trace metals in soil, water, air, etc.
Ion Liquid Chromatograph	Used to determine anion concentrations in drinking water, wastewater, and hydrogen peroxide Perchlorate determination

INSTRUMENT	FUNCTION
(2) High Pressure Liquid Chromatography systems which utilize (3) different technologies of detectors; UV detector, fluorescence detector, and a photodiode array detector	Pigment concentrations Analysis for explosives Herbicide analysis; diquat and paraquat Pesticide analysis; glyphosate Carbamates
Mercury Analyzer	Low level mercury in water/food; ppb level analysis
Flow-injection Ion Analyzer	Total N_2, ammonia
UV/VIS Spectrophotometer	Quantitative Filter technique applications for the determination of pigment concentrations present in suspended particles in ocean water Pigment concentrations of dissolved organic matter in ocean water Pigment concentrations in plant life
Pensky-Martens Closed Flash Point Tester	Flash Point determination of wastewater

Additional general support equipment includes balances, autoclaves, centrifuges, pH meters, turbidity meters, dissolved-oxygen (DO) meters, temperature and specific conductivity probes, total suspended particulate monitors, particulate matter of 10-m material (PM-10) monitors, composite surface-water samplers, and ground-water monitoring equipment.

5.3.1.1.3   Regulatory Authorities

All analyses performed by the Environmental Laboratory conform to strict requirements of:

- Comprehensive Environmental Response, Compensation and Liability Act (CERCLA)
- National Polluting Discharge Elimination Systems (NPDES)
- National Primary Drinking Water Regulations (NPDWR)
- Resource Conservation and Recovery Act (RCRA)
- Safe Drinking Water Act (SDWA)
- Clean Water Act (CWA)
- Clean Air Act (CAA)

5.3.1.1.4   Analytical Parameters and Methods

Analytical Parameters	Method	Analytical Parameters	Method
Biochemical Oxygen Demand	405.1	Carbamates	531.1
Chemical Oxygen Demand	410.4	Glyphosate	547
Oil and Grease	1664	Diquat	549.1
Settleable Solids	160.5	TPH	8015
Total Suspended Solids	160.2	TCL-Pesticide/PCBs(P/PCBs)	8081
Total Dissolved Solids	160.1	Freon	502.1
ICAP Metals	200.7	Volatile Organic Compounds	502.2
GFAA Metals	200.9	Volatile Organic Compounds MassSpec	524.2
Gas Generator Metals	245.1	TCL-VOC	8260b
TAL-Metals(23)	200.7 CLP	TCL-SVOC	8270c
Total Nitrogen(TKN)	351.3	Acid Extractable Fraction MassSpec	625/8270
Ammonia	350.3	Base Neutral Fraction(MS)	625/8270
Total Phosphorus	365.3	Residual Chlorine	330.5
Ortho Phosphorus	300.0a	Dissolved Oxygen	360.1
Anions(10)	300a&b	Temperature	
Turbidity	180.1	pH	150.1
Cyanide	335.2	Conductivity	120.1
Fecal Coliform	SM9222D	Hardness	200.7
Total Coliform	SM9222B	Sites Sampled	
Total Organic Carbon	415.2	Flow Rates	
Microwave Digestion	*	PM-10 Air Samplers	
Hotplate Digestion		TCLP	1613
Phenols	420.1	Metals extraction	1613
Sept Funnel Extraction BNAE		Metals fraction analytical	1613
Organic sample concentration		Semi-volatile extraction	1613
Pesticides	505/8081	Semi-Volatile Fraction	1613
Herbicides	515/8150	ZHE extraction	1613

5.3.1.2   Gas and Material Analysis Laboratory

The Science Laboratory Services Gas and Material Analysis Laboratory (GMAL) supports NASA as well as other Government agencies and aerospace contractors. This branch is divided into two sections. The Gas Analysis section conducts general and non-routine analysis on gas and cryogenic propellants, pressurants and hydraulic fluids. The Materials Laboratory conducts failure analysis and evaluates contaminants on metallic and non-metallic materials. Both labs provide investigative efforts necessary to understand and solve gas and material problems associated with rocket engine testing, plume analysis, and material compatibility issues by performing in depth lab and field studies.

5.3.1.2.1   Gas Analysis

The analytical instrumentation supporting the gas analysis capability is important in meeting the stringent NASA specifications required for the SSC propulsion testing programs.

- Several specialized Gas Chromatographs (GCs) capable of separating a gas sample into its individual components and identifying the components by comparison to a known certified standard. The GCs are used to measure impurities in He and H_2.

- Non Dispersive Infrared process analyzers provide a fast, accurate method for CO/CO_2 concentration analyses in gasses.

- Paramagnetic analyzers are used to determine oxygen impurities in N_2 and He gases and are used to certify Liquid Oxygen.

- Four phosphorous pentoxide analyzers are used to determine the moisture concentration in a variety of gases in the low parts-per-million range.

5.3.1.2.2   Material Analysis

The Materials Lab provides investigative and evaluative support for failure analysis and for identification of unknown materials and contaminants. In addition, the lab performs surface feature characterizations for fractures and flaws in flight and ground support equipment.

- Scanning Electron Microscopes (SEM) acquires a 3D image to reveal fractures, inclusions and to perform physical measurements. In addition, the SEM is used to establish evidence of stress and/or corrosion. This often allows for identification of the source, such as chemical reactions or mechanical wear. The SEMs are equipped with an energy-dispersive x-ray spectrometer, allowing for qualitative analysis. X-ray surface mapping can be performed to display the arrangement and distribution of elements and is useful to assess the phase composition of alloys or sample homogeneity.

- Fourier Transform Infrared (FTIR) spectrometers identify and quantify a chemical compound according to its molecular spectra. The FTIRs are used to identify NVR-inclusive hydrocarbon/non-hydrocarbon contaminants such as elastomers and softgoods.

GMAL's Major Analytical Instrumentation and Diagnostic Functions.

INSTRUMENT	FUNCTION
Gas Chromatograph (4)	Helium and Hydrogen gas analysis
FTIR (2)	Nonvolatile residue (NVR) analysis
FTIR / ATR	Non metallic material identification
Hydrocarbon Analyzers (2)	Hydrocarbon as methane concentrations in gases
Moisture Analyzers (4)	Moisture concentrations in gases
CO/CO_2 Analyzer (1 each)	CO/CO_2 concentration in gases
Oxygen (O_2) Analyzer (2)	Measures percent oxygen (O_2) in Gases
Trace Oxygen (O_2) Analyzer (1)	Impurities in N_2 and He
Laminar Flowbench	Particulate analysis and identification
Radionuclide Analyzer	Field analysis requiring qualitative and quantitative analysis of materials
Scanning Electron Microscope (2)	Micro-contamination analysis
FTIR Microscope (1)	Organic contamination analysis

Gas and Material Laboratory Capabilities

Tasks	Tasks
THC	CO/CO_2
Moisture	GC / Helium
Particulates (Gas @ Range)	GC / Hydrogen
Particulates (Fluids @ Range)	GC / Oxygen
O_2 Purity	pH
Trace O_2	Conductivity
Coulometric Titrations	Condensable Hydrocarbons
Fe Hydraulic Fluid	Acetylene Content
CFC's Hydraulic Fluid	Scientific Support
% H_2O_2	H_2O_2 Passivation
Ammonium	Failure Analysis
Hydrocarcon / NVR	Micro-contamination

5.3.2 Measurement Standards and Calibration Laboratory (MS&CL)

The MS&CL performs calibrations, test and repair services. These calibrations and measurements are all traceable to National or Intrinsic standards of measurement. Repair services vary from basic on mechanical instruments to component level on electrical instruments. The laboratory actively participates in the NASA Metrology and Calibration Working Group and maintains membership in the National Conference of Standards Laboratories (NCSL). Laboratory Capabilities are presented in Table 5.3.2-1.

5.3.2.1   Standards Laboratory

The Standards Laboratory transfers extremely accurate electrical and physical calibrations from reference standards to secondary transfer and working standards. These secondary standards are then used in measurements and other calibrations. Many standards laboratory measurements utilize statistical techniques designed to reduce measurement uncertainty. The accuracy and uncertainty requirements found in the standards laboratory result in many accuracy ratios that are less than 4:1. The laboratory participates in NASA Measurement Assurance Programs (MAPs) that help to ensure laboratory capabilities. These MAPs include Volt MAP, Resistance MAP, Extended Resistance MAP, Mass MAP, Gage Block MAP, Accelerometer MAP and Temperature MAP. In addition the laboratory utilizes the NASA Portable Josephson Array Voltage Standard to maintain DC voltage traceability. Highlights of the Standards Laboratory capabilities include:

- DC Voltage: 0 to 1000 volts
- AC Voltage: 0 to 1000 volts
- Resistance: 0.0001 Ω to 100 MΩ
- Mass: 1 mg to 60 kg
- Pressure: 0.2 to 12,140 psig

5.3.2.2   Dimensional Laboratory

The Dimensional Laboratory performs calibrations and measurements on geometrical characteristics of instruments and components. The Laboratory serves the NASA metrology community actively as the pivot laboratory for the NASA Gage Block MAP. Laboratory Capabilities include:

- Lengths: 0 to 80 inches
- Roundness: 0 to 14 inches diameter
- Threads: 4 to 80 pitch
- Surface Plates: 0 to 1,000 sec
- Optical Angle: 0° to 360°
- Surface Roughness: 20 μin to 125 μin
- Hardness: accuracy of ± 1.0 unit of the Rockwell Standard

5.3.2.3   Electrical Calibration and Instrumentation Repair Laboratory

The Electrical Calibration Laboratory provides calibration and repair services for a wide variety of general measurement equipment. The laboratory is equipped with automated systems that provide automatic and semiautomatic calibrations. The following is a sample of laboratory capabilities:

- Multimeters: Volt (0 to 1000 v), Ohm (1 to 1 GΩ), Current (0 to 100 A)
- Oscilloscopes: 50 μv/Div to 5 v/Div into 50 Ω, 50 μv/Div to 200 v/Div into 1MΩ, 5 sec to 0.4 nsec
- Signal Generators: 0 to 18 GHz, 0 to −172 db
- Sound-Level: 20 Hz to 2.5 kHz
- Frequency Counter Time Base: ± 5 x 10^{-12}
- Watt-meters: 0 to 50 W, 10 to 1000 MHz
- Tachometers: 15 to 20000 rpm
- Photometers: 0.1 to 1000 foot-candles
- Radiation Survey Meters: less than 5,000 mR
- Phase Angle: 0 to 360°
- Combustible Gas Monitors: 0 to 100% LEL
- Oxygen Level Meters: 0 to 21.9% O_2

The instrumentation section of the laboratory performs modifications, rework, repairs and functional checks of all types of instrumentation, amplifiers and signal conditioners used for testing and data collection as well as other miscellaneous types of electronic equipment. The following types of instrumentation and equipment are a representative sample of overall capabilities:

- AC amplifiers and DC amplifiers (both manual and computer programmable), transducer amplifiers, signal conditioners, and filter modules.
- High intensity light sources, chart recorders, regulated and unregulated power supplies.

5.3.2.4   Pressure Laboratory

The Pressure Laboratory provides calibration and repair of pressure measuring and pressure generating equipment such as transducers, digital and analog gages, and dead weight testers. The laboratory provides in-place calibration services throughout SSC. Calibration services can be scheduled as to maintain the cleanliness level of instrumentation under test. Automated pressure calibrations are available to 60,000 psig. Some capabilities of the pressure laboratory are listed on the next page:

- Calibration using air and hydraulics to 60,000 psig
- Vacuum calibration down to 1 mm of mercury (mmHg)

5.3.2.5　　Flow Laboratory

Calibration services are provided for liquid and gas flow measuring instruments. Direct air and nitrogen gas calibrations are available. Flow characteristics of other gases, such as hydrogen or helium, can be computed:

- Liquid flow from 0.01 to 300 gal/min
- Gas flow from 0 to 800 cfm
- Moisture analysis from 0.1 to 1,000 ppm

5.3.2.6　　Mechanical Laboratory

Calibration is provided for various force measuring instruments. Capabilities include:

- Torque calibrations from 0 to 2,000 ft-lbs
- Force calibrations from 0 to 120,000 lbf

5.3.2.7　　Wind Tunnel

Calibration of air velocity instruments and large airflow devices are performed in the Wind Tunnel's 3 ft. diameter test area. Capabilities range from 0 to 80 m/s.

5.3.2.8　　Vibration Laboratory

The Vibration Laboratory calibrates accelerometers and other vibration measuring equipment.

5.3.2.9　　Temperature Laboratory

The SSC Temperature Laboratory is an active participant in the NASA Temperature Measurement Assurance Program (TMAP). Participation includes the additional responsibilities of pivot laboratory. As pivot laboratory the SSC lab serves the NASA metrology community as the coordinator of all TMAP activities and as a metrology resource for all temperature measurement activities.

The Temperature Laboratory performs calibrations on all types of temperature and humidity sensing devices. Laboratory capabilities include:

- Temperature calibration from −180 to 500 °C.
- ITS90 fixed point calibrations at the Triple Point of Mercury, the Triple Point of Water, the Freezing Point of Tin and the Freezing Point of Zinc.
- Temperature and Humidity chamber: Relative Humidity from 15 to 95%.

5.3.2.10 Cleaning Laboratory

The Cleaning Laboratory performs cleaning and certification of instruments to NASA Standard 79-001 for LOX service. This certifies that the hydrocarbon content and particulate size and count are within the constraints of the standard.

5.3.2.11 Metrology Engineering

Metrology Engineering provides technical support concerning metrology and calibration problems. Specifically it provides:

- Support to the laboratories when non-routine and/or difficult measurements or calibrations occur. This involves assuring the proper equipment and techniques are used for each measurement. When such non-routine measurements are made, the metrology engineering staff will establish total uncertainty to assure proper accuracy in assignment to the test items.

- Support to the customer during and after calibration or measurements by acting as a liaison between the customer and the technicians who performed the calibration to ensure that all requirements are understood and provided. In the event post calibration problems occur the metrology engineering staff will provide support to the user in the evaluation of the problem.

- Conducts and acts as lead on all SSC Measurement Assurance Programs (MAPs).

- Serves as MS&CL technical resource by remaining current in the latest metrology and calibration theories, equipment, methodologies, research and standards. Serves the laboratory as member delegate to the National Conference of Standards Laboratories (NCSL). Participates and actively supports the NASA Metrology and Calibration Working Group.

- Metrology Engineering is also responsible for other laboratory functions such as reverse traceability, calibration interval assignment and adjustment, and laboratory point of contact for all auditors.

- Provides specific support as required for any special and/or unusual technical problems for the lab.

Table 5.3.2-1 MS&CL Capabilities

PARAMETER	RANGE	BEST UNCERTAINTY	STANDARD
DC VOLTAGE	0.1 V	± 1.1 ppm	Solid State Voltage Reference and Precision Divider
	1.0 V	± 0.5 ppm	Datron 4910 Solid State Voltage Reference
	1.018 V	± 0.5 ppm	Datron 4910 Solid State Voltage Reference
	10 V	± 0.5 ppm	Datron 4910 Solid State Voltage Reference
	100 V	± 1.1 ppm	Solid State Voltage Reference and Precision Divider
	1000 V	± 1.1 ppm	Solid State Voltage Reference and Precision Divider
AC VOLTAGE	2 mV to 0.6 V @10 Hz to 1 MHz	± 6 to 2900 ppm	Fluke 792A Thermal Transfer Standard
	1 V to 30 V @10 Hz to 1 MHz	± 18 to 108 ppm	Holt 11 Thermal Transfer Standard
	10 V to 1000 V @10 Hz to 100 kHz	± 17 to 110 ppm	Holt 11 Thermal Transfer Standard
RATIO, DC	$1:1 \times 10^{-7}$ to 1:1.1 to 1:1	± 0.1 ppm	Ratio Build-Up Terminal Resistance Boxes
RATIO, AC	1:001 to 1:1 @ 1 kHz	± 0.5 ppm	ESI DT72A Ratio Transformer
RESISTANCE	0.0001 Ω	± 5.0 ppm	Standard Resistors, Standard Current Source and DVM
	0.001 Ω	± 5.0 ppm	Standard Resistors, Standard Current Source and DVM
	0.01 Ω	± 2.5 ppm	Standard Resistors, Standard Current Source and DVM
	0.1 Ω	± 2.5 ppm	Standard Resistors and Guildline Current Comparator
	1 Ω	± 1.0 ppm	Thomas 1 Ω Standard Resistor
	10 Ω	± 2.5 ppm	Standard Resistors and Guildline Current Comparator
	100 Ω	± 2.5 ppm	Standard Resistors and Guildline Current Comparator
	1 kΩ	± 2.5 ppm	Standard Resistors and ESI Resistance Bridge

PARAMETER	RANGE	BEST UNCERTAINTY	STANDARD
RESISTANCE (continued)			
	10 kΩ	± 1.0 ppm	ESI SR104 Standard Resistor
	100 kΩ	± 2.5 ppm	Standard Resistors and ESI Resistance Bridge
	1 MΩ	± 3.0 ppm	L&N 4050-B Standard Resistor
	10 and 100 MΩ	± 5.0 ppm	Standard Resistors and DMM
	> 100 MΩ	± 0.05% to 1.0%	Guildline TeraOhm Meter
DC CURRENT	20 A to 100 A	± 0.05%	Valhalla 2575A Current Shunt
	2 A to 20 A	± 25 ppm	Fluke Y5020 Current Shunt
	1 μA to 2 A	± 10 ppm	Standard Resistors and DVM
AC CURRENT	10 A to 100 A DC to 1 kHz	± 0.1 %	Valhalla 2575A Current Shunt
	2 A to 20 A 1kHz to 10 kHz	± 25 ppm	Fluke Y5020 Current Shunt
	10 mA to 2 A 10 Hz to 5 kHz	± 50 ppm	Fluke A40 Current Shunts
CAPACITANCE	1 pF to 1.111μF	± 10 ppm @1 kHz	Andeen Hagerling 2500A
	100 pF	± 20 ppm @100 Hz ± 10 ppm @1 kHz	GR 1404B Standard Capacitor
	1000 pF	± 10 ppm @100 Hz ± 20 ppm @1 kHz	GR 1404A Standard Capacitor
INDUCTANCE	0.1 mH to 99,999 H	± 0.12% @100 Hz and 1 kHz	GR 1689M Digibridge
	100 μH	± 0.167% @100 Hz ± 0.152% @1 kHz	GR 1482-B Standard Inductor
	1 mH	± 0.035% @100 Hz ± 0.038% @1 kHz	GR 1482-E Standard Inductor
	10 mH	± 0.032% @100 Hz ± 0.032% @1 kHz	GR 1482-H Standard Inductor
	100 mH	± 0.032% @100 Hz ± 0.033% @1 kHz	GR 1482-L Standard Inductor
	1 H	± 0.030% @100 Hz ± 0.080% @1 kHz	GR 1482-P Standard Inductor
	10 H	± 0.032% @100 Hz ± 0.032% @1 kHz	GR 1482-T Standard Inductor

PARAMETER	RANGE	BEST UNCERTAINTY	STANDARD
RF ATTENUATION	0 to -120 db 10 MHz to 18 GHz	± 0.02 db	NIST certified attenuator and NIST traceable measurement receiver.
RF POWER	0 to 50 W 10 to 1000 MHz	± 0.5% of reading ± 0.1 Hz	NIST certified sensor, frequency counter and power signal source.
FREQUENCY TIME BASES	100 kHz, 1 MHz, 5 MHz & 10 MHz	Accuracy: $\pm 5 \times 10^{-12}$ Stability: $\pm 2 \times 10^{-12}$	Cesium Beam frequency standard representing a natural physical constant.
ANGULAR FREQUENCY (RPM)	15 to 20,000 rpm	± 0.3% range	Tachometer calibrator traceable to a natural physical constant.
PHASE ANGLE	0 to 360°	± 0.01°	NIST traceable phase angle standard.
Dimensional **LENGTH**	 0.05 to 20 inches	 ± 1 μin to ± 4 μin	 Gage Blocks and Mechanical Gage Block Comparator.
	0.5 to 100 mm	± 0.05 μm to ± 0.06 μm	Gage Blocks and Mechanical Gage Block Comparator.
	0 to 80 inches (0 to 2.032 m)	± 50 μin absolute or ± 20 μin when used with corrections	Gage blocks and a supermic/80 inch measuring machine.
INSIDE MEASUREMENT	0.25 to 14 inches	± 10 μin	Internal Super Micrometer
PITCH DIAMETER	4 to 80 pitch	± 10 μin	Thread Wires and a Super Micrometer
SHADOW AND PROFILE PROJECTION	0 to 14 inches	± 0.0001 inch	Gage Blocks and an Optical Comparator
OPTICAL ANGLE	0° to 90° in 1 second steps	± 0.18 arc seconds	Angle Blocks
	0° to 360° in 1° steps	± 0.14 arc seconds	Indexing Table
FLATNESS	0.05 to 6 inches	± 1 μin	Interferometer
(SURFACE PLATES)	0.0 to 1000 sec	± 4 and ± 20 seconds	Federal Surface Plate Calibrator
COATING THICKNESS	6.3 microns to 1.75 mm	± 5%	NIST Standard Reference Materials

PARAMETER	RANGE	BEST UNCERTAINTY	STANDARD
SURFACE FINISH	20 μin to 125 μin	±2 μin to ± 6 μin	Federal Surface Analyzer
HARDNESS	ROCKWELL B&C Scales	± 1.0 unit of the ROCKWELL Std.	Industrial ROCKWELL hardness standards and hardness comparator
ACOUSTICS Sound Pressure Level	114 db from 20 Hz to 2.5 kHz	± 0.2 db	Microphone reciprocator traceable to a natural physical constant for frequency and to NIST for sound level.
TEMPERATURE	-50°C to 300°C	± 0.01° C	Platinum Resistance Thermometer with Rosemount Oil Bath
	0.01° C (fixed)	± 0.0005° C	Triple Point of Water Cell
	-38.8344° C (fixed)	± 0.0005° C	Triple Point of Mercury Cell
	231.928° C (fixed)	± 0.005° C	Freezing Point of Tin Cell
	419.527° C (fixed)	± 0.010° C	Freezing point of Zinc Cell
PRESSURE	0.2 to 600 psia/psig	± 50 ppm	Ruska 2465 Dead Weight Piston Gage
	6 to 12,140 psig	± 100 ppm	Ruska 2400HL Hydraulic Dead Weight Piston Gage
	30 to 60,000 psig	± 100 ppm	DH 53100 Hydraulic Dead Weight Piston Gage
VACUUM	0.001 to 1 Torr	From ± 2.53% @ 0.001 Torr to ± 0.70% @ 1 Torr	MKS 390HA Capacitance Vacuum Standard
MASS	1 mg to 60 kg	± 10 μg to 100 mg	Mass Standards and Balances
FORCE	0 to 600 lbf	± 14.3 lbf	Dead Weight
	5000 lbf	± 0.59 lbf	Morehouse Proving Ring
	10,000 lbf	± 0.81 lbf	Morehouse Proving Ring
	60,000 lbf	± 6.4 lbf	Morehouse Proving Ring
	120,000 lbf	± 24 lbf	Morehouse Proving Ring
TORQUE	0.5 to 215 ozf·in	± 0.2% reading	Waters Torque Calibrator
	10 lbf·in to 2,000 lbf·ft	± 0.1% reading	AKO Torque Calibrators

PARAMETER	RANGE	BEST UNCERTAINTY	STANDARD
Flow			
LIQUID	0.01 to 300 gals/min	± 0.1% reading	Cox 311 Volumetric Flow Calibrator
GAS	0 to 50,000 cm^3	± 0.2% reading	Sierra Cal Bench
	0 to 20 ft^3/min	± 0.35% reading	Cubic Foot Bottle
	0 to 800 ft^3/min	± 0.58% reading	Sonic Nozzles
WIND SPEED	0 to 1000 fpm	± 10 fpm	TSI Hot Wire Anemometer
	1000 to 15,748 fpm	± 1% of reading	Standard Pitot Static Tube
MOISTURE	0.1 to 1000 ppm	± 0.5 ppm	Standard Dew Point Hygrometer
GAS METER CALIBRATION			
Hydrogen & LEL Meters	0% H2(0.0% LEL) 0.4% H$_2$(10% LEL) 1.0% H$_2$(25% LEL) 2.0% H$_2$(50% LEL)	≤ 0.001%(0.025% LEL) ± 1%(0.1% LEL) ± 1%(0.25% LEL) ± 1%(0.5% LEL)	Certified Ultra-High Purity Helium Certified Hydrogen/Air Mixture Certified Hydrogen/Air Mixture Certified Hydrogen/Air Mixture
Oxygen Monitors	0% 10.4% 20.8%	≤ 0.01% O$_2$ content ± 0.21% O$_2$ content Intrinsic Standard	Certified Ultra-High Purity Nitrogen Certified Helium/Air Mixture Ambient Air free of interference gases
Carbon Monoxide Monitors	10 ppm CO 20 ppm CO 1430 ppm CO	± 0.2 ppm CO ± 0.4 ppm CO ± 14.3 ppm CO	Certified Carbon Monoxide/Air Mixture Certified Carbon Monoxide/Air Mixture Certified Carbon Monoxide/Air Mixture
Carbon Dioxide Monitors	3% CO$_2$	± 0.06% CO$_2$	Certified Carbon Dioxide/Nitrogen Mixture
Sulfur Dioxide & Hydrogen Peroxide Meters	10 ppm SO$_2$ or H$_2$O$_2$	± 0.2 ppm SO$_2$ or H$_2$O$_2$	Certified Sulfur Dioxide/Nitrogen Mixture
HUMIDITY			
RELATIVE	15% to 95%	± 1% Relative Humidity	Standard Psychrometer System
ABSOLUTE	-100°F to 167°F	± 0.36° F	Standard Dew Point Hygrometer.
OPTICAL RADIATION / PHOTOMETRY	0.1 to 1000 foot-candles	± 2.5%	NIST traceable incandescent lamp and current source.

PARAMETER	RANGE	BEST UNCERTAINTY	STANDARD
VIBRATION	10 Hz to 10 kHz @ 10 g pk	10 to 50 Hz ± 2%, 50 Hz to 2 kHz ± 1%, 2.0 kHz to 10 kHz ± 2%	Bruel and Kjaer 8305 Standard Accelerometer
FLUID DENSITY/ SPECIFIC GRAVITY	0.001 to 2.0 g/cm^3	± 0.00001 g/cm^3	Mettler DMA58 Density Meter
MAGNETICS, AXIAL & TRANSVERSE	311 to 10,000 gauss	± 3%	Standard Magnets
VOLUME, LIQUID	0.001 ft^3 to 1 ft^3	± 2.2 x 10^{-7} to ± 0.024 ft^3	Volume Computed with Density and Mass of Water
PULSE RISETIME	≤ 70 pSec	± 6pS	NIST traceable pulse generator.

5.3.3 Prototype Development Laboratory

The Prototype Development Laboratory offers electronic and mechanical prototype development. The laboratory offers a kind of one-stop shopping for its customers, adding an element of convenience to systems development. Staffing consists of technicians with a wide variety of expertise ranging from microprocessor development to undersea systems development, field support, and precision machining. Engineering staff is available for additional software and hardware development and support. This provides the customer with a single point of contact for project development or technical support.

Mechanical fabrication services are available for a wide assortment of project support, including mill and lathe work, welding and sheet metal. The lab posses in-depth experience in working with aluminum, stainless steel, copper, brass, delrin, PVC, lexan, and many other metal and plastic material stocks.

The laboratory provides in-house printed circuit board design and fabrication of boards up to 11 in. x 17 in. double sided. Typical turn-around time is less than 48 hours depending on the complexity of the board layout.

Electronic fabrication services are available for design and integration of electronic systems, ranging from design and layout to integration of off-the-shelf components into custom enclosures. A wide variety of test equipment is available including oscilloscopes, logic analyzers, signal generators, counters and power supplies.

The Prototype Development Laboratory is the only group at SSC with the unique capability to marry mechanical and electronic systems together to form completed turn-key systems ready for field testing and deployment.

Laboratory capabilities are summarized below:

Technical Expertise
- Electronic Support
- Analog and digital systems development
- Fabrication of electronic systems and subsystems
- Electronic packaging and integration
- PC board fabrication, designs, layouts and wire wrapping
- Prototype development and field testing
- Repairs and upgrades of electronic systems
- Systems field testing, evaluation, and support

Field Support
- At sea support for the deployment and operation of oceanographic systems
- Array winch refurbishment and operation
- Field installations and repairs of electronic systems

General Support and Services
- Equipment market surveys
- Component/material procurement (quick response)
- Weekly project financial reporting
- Provide technical support in customer facilities

Mechanical Support
- Machining of small assemblies
- Refurbishment of mechanical equipment (winches, pumps)
- Fabrication of mechanical equipment with electronic controls
- Fabrication of replacement parts

Laboratory Equipment Electronic Test Equipment
- Logic Analyzer (138 channel acq., 48 channel stimulus)
- Analog and Digital Oscilloscopes (up to GHz bandwidth)
- PROM, EPROM, EEPROM, Microcontroller Programmer
- Frequency generators and counters
- High resolution DVOMs
- Precision voltage and current sources
- PC Board Fabrication Machine

5.4   Information Systems

Information Systems provides data acquisition and control services for all SSC facilities and resident agencies. The TTSC provides the following categories of support:

- Applications Support and Software Development
- Data Systems and IT Operations
- Telecommunications
- IT Security Services
- Information Resource Management
- Process Engineering
- Planning & Scheduling Services
- Audio/Visual and Video Services

Support is provided to NASA and NASA contractors, Resident Agencies, universities and commercial entities under the direction of NASA points of contact for shared services and specific task orders. A variety of computing architectures are supported, including supercomputer, mainframe, client-server, web-based thin clients, and desktops.

The above list of support categories are described in the following paragraphs.

5.4.1 Applications Systems

Applications Support encompasses life-cycle support of hardware and software systems for business, scientific and engineering applications.

Business support focuses on maintenance of Financial and Institutional applications software, ranging in scope from web-based applications, desktop client-server, to mainframe applications.

The SSC Mainframe LPAR is located at Marshall Space Flight Center (MSFC) and supports the financial, procurement, shipping and receiving systems for NASA and its contractors. It also hosts several NASA Agency-wide applications for the NASA user community. The Stennis LPAR receives its system programming support from the NASA ADP Consolidated Computer Center (NACC) at MSFC but local applications programming support is provided.

Institutional applications are increasingly developed to utilize the ubiquitous World Wide Web (WWW) browser software provided on every desktop computer. This approach allows familiar, standardized access to a wide range of critical functions, including financial reporting, administrative, and programmatic activities.

Scientific support is also provided for applications software.   In addition, software engineering and systems integration/implementation/ administration is performed. Areas of expertise include: satellite applications, systems analysis, data visualization/computer graphics/image processing, and satellite applications. Software development and maintenance are also performed using DOD's Major Shared Resource Center, consisting of Cray and Silicon Graphics supercomputers, which are resident at SSC as part of the Naval Oceanographic Office (NAVOCEANO).

Direct support is provided to the Propulsion Test Directorate for engineering and business needs, such as data acquisition, web portal, process automation, information management, and data warehouse/mining capabilities. Integrating commercial off-the-shelf applications and database management systems such as Oracle and SQL Server to solve propulsion test data problems is a major component of this support.

5.4.2 Software Development

Software development is performed for applications to be run on desktop, web-based, client-server, GIS, mainframe, automated workflow processes. Software development practices follow the best-of-breed methodologies developed by the Software Engineering Institute for their Software Development Capability Maturity Model (SW-CMM).

Development and operational languages include: C, C++, COBOL, JAVA, FORTRAN, Visual Basic, VBA, LabView, HTML, and Assembly. COTS development environments include: Oracle, MatLab, PV-Wave, Arc/Info, X Motif, Open GL, Windows API, Embedded SQL, Power Builder, Natural, Informed, Windchill, and Ultimus. Scripting languages include: Perl, C shell, Bourne shell, Expect, Tcl/Tk, JCL, and REX. Web development tools include: Front page, ArcIMS, Dreamweaver, ASP, and Visual Interdev.

Problem domains addressed include: web-based e-business, financial, database, engineering/scientific, remote sensing/GIS, real-time, data acquisition, data visualization and modeling, data fusion/integration.

5.4.3 Data Systems and IT Operations

Data Systems and Operations support encompasses operation and administration of the Stennis Data Center, the SSC IT Security Program, NASA-owned telecommunications assets, and IT planning.

Application and web page hosting, pooled license management, production data processing and digitization, CD mastering services, high-end print capability, Tape/CD library management, backup, archival and restoration services are provided. Capability and services include:

- 24X7 operational capability
- secure 4 tier network environment
- redundant power systems
- high performance network connectivity

- NT, Linux, HP-UX, Irix, Solaris computing environments
- ADABASE, Oracle, SQL Server databases
- TCO, Capacity, consolidation and obsolescence planning
- Disaster recovery and COOP planning
- Systems administration
- Database administration
- Website registration and administration
- Performance tuning and monitoring

The Stennis Data Center (commonly referred to as PSCS) is evolving from what was once primarily a financial and administrative computer room into a full-function data center supporting e-business as well as unique Propulsion Testing IT requirements. Data Operations provides a wide range of computer systems that are critical for the Stennis Space Center community. The STDC hosts web-based and client server applications and remote Mainframe computer applications for the Stennis user community.

5.4.4 Administrative Telecommunications Support

The TTSC provides administrative support for critical NASA-owned telecommunications assets and data services. Administrative support is provided for:

- Data Center servers
- Non-ODIN desktops
- Coordination of WAN off-site circuits
- Pager services
- Video-teleconferencing facilities
- Billing reconciliation for ODIN, NISN, telephones
- Reimbursable billing reports development

5.4.5 IT Security Services

Information technology security is an integral part of all IT development and operations, ensuring ongoing access, availability, and integrity of critical systems and information assets. The TTSC contract provides staffing and expertise to the NASA IT Security Manager for development, administration and execution of the IT Security Program at Stennis. While the NASA IT Security Program is coordinated at the Agency-level, it is administered locally at each Center, following guidance provided by NPG 2810.1. Coordination is also maintained with onsite Resident Agencies to ensure compatible and complementary programs. Capabilities and services include:

- IT Security plans and Contingency plans
- Intrusion detection scanning and analysis
- Incident response and investigations
- Risk analyses

- System Vulnerability assessment
- Critical infrastructure planning
- Security notifications and bulletins
- Security awareness training

5.4.6 Information Resources Management

The TTSC develops and administers systems to assist SSC users in the effective management of critical information. Categories of information resource management include:

- Documentation management systems
- Drawing management systems
- Image management systems
- Records management systems
- Knowledge management systems

5.4.7 Process Engineering

Process engineering capabilities not only provide for effective operations and production, but also the continuing assurance of quality output and products. Services include:

- Process definition and analysis
- Design and re-engineering
- Workflow automation
- Configuration management
- Continuous quality improvement
- Certified ISO audits

5.4.8 Planning and Scheduling Services

Strategic and tactical IT planning support is provided through the TTSC contract to the NASA/SSC Chief Information Officer and Center Services Chief. Programmatic and institutional requirements are mapped to existing IT assets and infrastructure. Alignment with NASA Agency, as well as relevant industry standards, is ensured. Annual and five year technology budget and technical planning is supported. Planning and scheduling services include:

- IT strategic and tactical planning
- Annual and five year equipment and technology plans
- Project planning, tracking, reporting,
- Cost, schedule, resource allocation, forecasting
- Scheduling tools

The Scheduling group delivers integrated, scalable project management solutions for the Propulsion Test and Information Systems Directorates. The Group is responsible for the coordination of workflow within or between work areas and units, the development and/or review of program schedules, work documents, and daily work area priorities.

5.4.9 Audio-Visual and Video Services

Audio-visual and video services are available to support all types of conferencing, multi-media and presentation requirements. Capabilities include:

- Conference room event support
- Conference room multimedia design and configuration
- ViTS reservations/facilitation
- A/V equipment loaner services
- Multimedia event setup and production
- Analog and digital video production in multiple formats
- Closed captioning support
- Test firings, documentaries, training, safety, special events

6.0 FACILITY OPERATING SERVICES (FOS)

FOS, provided by an onsite contractor to NASA, supports NASA and the resident agencies at SSC. Areas of support, as related to the SSC propulsion-testing mission, include Facility and Operations Engineering Services, Facility Services, Institutional Services, and limited Technical Services. Additional information related to these select facility and operations support services is contained in the following sections.

6.1 Facility and Operations Engineering Services

The FOS Contractor provides engineering services in the planning, project development, design, construction, modification, repair, and maintenance of site facilities, systems, and equipment. Also, FOS Engineering assists with documentation, energy management, space utilization, spares provisioning, and certification of pressure vessels.

6.1.1 Engineering Services

This service provides the core engineering support such as design, specification development, drawings, construction monitoring and evaluation of standards and requirements compliance. Engineering evaluations, studies, and reports are provided on a project as required. The FOS contractor is supplemented by subcontracting with A&E firms as required for specific engineering talent not available at SSC. Local manpower agencies also provide for supplemental engineering talent.

6.1.2 Documentation

Full documentation services are provided, including the monitoring of all project documentation and the writing and revising of Maintenance Instructions, SSC Standards, and Technical Procedures. The Computer Aided Design and Drafting (CADD) operators prepare engineering drawings on the modern CADD equipment, utilizing both Computer Vision and AutoCAD software systems. Central Engineering Files (CEF) maintains files on all historical and current engineering documentation. Copies of any documentation are readily available to requestors through CEF.

6.1.3 Energy Management and Space Utilization

The Energy Management and Space Utilization function provides specific information regarding energy costs, utilization, personnel headcount within buildings, and configuration changes to the buildings.

6.1.4 Spares Provisioning and Certification of Pressure Vessels

The Spares Provisioning program monitors the availability of spare components and parts used in the Test Complex area. Through this program, FOS Engineering ensures that spares are either available in the warehouse or procured when needed. Also for the Test Complex, FOS Engineering provides American Society of Mechanical Engineers (ASME) certification of existing pressure vessels and assists in the procurement, design, and certification of new pressure vessels. The FOS contractor currently holds both the R and U stamps, thereby being able to do both repairs and manufacture pressure vessels in accordance with ASME codes.

6.1.5 Construction Services

The FOS engineers are responsible for the implementation of construction projects through subcontractors. In doing so, they provide surveillance of the construction to ensure contract compliance for NASA and the resident agencies. As the principle liaison between the construction contractor and the Government, they are responsible for all aspects of the construction contract, including coordination, progressive payments, labor compliance, submittal approval, safety, and the closeout of their projects.

6.2 Facility Services

Facility services include numerous shops and operations; these are described in further detail in the following sections.

6.2.1 Carpentry Shop

The Carpentry Shop provides all carpentry services, from framing buildings to cabinet work, including the following: scaffolding, roofing, masonry, doors and door-closers, vaults, floor and ceiling tiles, partitions, furniture repair, etc.

6.2.2 Maintenance Engineering

Engineers oversee the maintenance of facilities, systems, and equipment, and conduct annual facility inspections to ensure that needed maintenance is identified. They provide predictive testing services, such as infrared, vibration analysis and oil analysis. Reliability centered maintenance studies systems to determine proper maintenance resource allocation. System engineering provides system surveillance and technical assistance to shop craftsmen.

6.2.3 Equipment Operations

Equipment Operations provides operators for mobile cranes, bulldozers, backhoes, forklifts, front-end loaders, dump trucks, derricks and road graders; and also performs equipment operator qualifications, rigging, and proof-testing.

A Mississippi state-certified sanitary landfill is maintained for daily disposal of nonhazardous wastes.

6.2.4 Equipment Maintenance

Equipment Maintenance personnel provide preventive and corrective maintenance (PM/CM) services for all types of gasoline- and diesel-powered machinery.

6.2.5 Preventive Maintenance Team

The Preventive Maintenance (PM) team performs PM on a wide variety of electrical and mechanical equipment. This composite shop services cranes, UPSs, elevators, derricks, jib and monorail hoists, bridge cranes, emergency lighting, MCCs, substation transformers, circuit breakers, fire alarm systems, the navigational lock and bridge, and cathodic protection.

6.2.6 Marine Operations

Marine Operations is responsible for operating a 1200 hp, push-type tugboat within the SSC canal system and the Mississippi/Louisiana coastal and inland waterways. They also handle the operation of SSC's navigational lock and bascule bridge.

6.2.7 Mechanical/Plumbing Shop

The Mechanical/Plumbing Shop personnel perform installation, modification, and corrective maintenance on plumbing/piping systems (with lines and valves up to a 30 in. diameter). They also carry out general mechanical activities on pumps, motors, lifting devices, test stand blast doors and windows, overhead rolling doors, and a wide variety of other mechanical systems. This shop also provides support to the Preventative Maintenance Shop in corrective maintenance of the navigational lock and bridge.

6.2.8 Electric Shop

The Electric Shop employees perform maintenance, repair, installation, and modification to all electrical systems and equipment, from 28 Vdc controls to the 13.8 kV high-voltage distribution system.

6.2.9 Heating, Ventilating, and Air-Conditioning Shop

The HVAC shop operates and maintains approximately 8,000 tons of air conditioning. Assets include 18 centrifugal and 30 reciprocating air and water cooled chillers ranging from 90 ton to 265 ton single systems. The shop also manages 750 main building air handling units, 90 computer room units, 525 heating/ventilation units, and 35 HVAC pneumatic control air compressors. Sixty low pressure hot water heating boiler systems consisting of approximately 165M BTUs of heating capacity are maintained by the

HVAC shop. These comprise both forced draft fire tube and natural draft water tube natural gas fired system and 5 geothermal heat lift machines.

6.2.10   Energy Management and Control System Office

The Energy Management and Control System Office (EMCS) provides centralized, 24-hr operation, control and monitoring of the HVAC equipment; sewer; potable water; natural gas; electrical; and other related facility, utility, and special-purpose systems in 55 SSC facilities. The office also prepares energy reports and analyses, manages fuel, and coordinates radio/communications and utility outages.

The EMCS consists of two separate control systems, the NASA Infoscan System and the Landis & GYR Powers System 600. NASA's Infoscan System consists of approximately 300 physical field points and 7 field stand-alone microprocessors in 7 buildings served by 4 networks support by a Hewlett Packard Series 1000A minicomputer work station. The Landis & GYR Powers System 600 incorporates approximately 10,000 physical field points and 130 field stand alone microprocessors in 45 main buildings served by 8 networks supported by a DEC PDP 11/93 as the central station minicomputer work station.

This office is also responsible for operation, maintenance, research, design, and development of centralized and stand-alone networked computer utility control systems.

6.2.11   Paint Shop

The Paint Shop provides all types of maintenance painting, including decorative and protective painting to interior and exterior building surfaces and equipment.

6.2.12   Roads and Grounds

Paved and unimproved roads, improved lawns and landscaping, unimproved land, parking areas, and surface drainage systems are maintained by Roads and Grounds personnel.

6.2.13   Drayage and Labor

Drayage and Labor provides pickup and delivery service, furniture moving, and general labor support to SSC organizations.

6.3   Institutional Support Services

The FOS contractor's Support Services Division (SSD) provides institutional support services. SSD provides NASA and resident agencies administrative support.

6.3.1 Logistics Services

6.3.1.1   Supply Operations/Inventory

Provides shipping/receiving, warehousing and storage, and inventory management control for material/equipment used by NASA and its' contractors. The main inventory contains 23,000 line items of general stores, program and standby stocks warehoused in buildings 2204 and 2203. Benchstock inventories are provided in buildings 2201, 2205, 2105, 4010 and 4302, which contain 14,900 line items of recurring demand items. Maintain and process Just in Time (JIT) contracts for non-powered hand tools and gas cylinders as well as tracking and validation for cylinder demurrage billing. Shipping/Receiving processes all incoming material and equipment at building 2204. The warehouse provides a Custodial Storage program for over 1,000 line items of special program items. Parcel delivery to all customers outside the test stands. Provides retrieval of used toner cartridges as well as paper delivery to the commonly used copiers within building 1100. Provides bonded storage within building 2204 as requested. Maintains shelf-life controls on DLA identified items. Provides all services within the ISO9001 guidelines. Provides outbound shipment via carriers and overnight services. Maintains the Fed-Ex Powership system for tracking of overnight packages. Provides tracking of hazardous shipments from manufacturer to SSC gate and on to requestor. Provides household moves as required.

6.3.1.2   Mail Services

Maintains and operates SSC facility post office in conjunction with the U. S. Postal Service for the processing of official mail; provides delivery/pickup to buildings each day; and hand-delivers registered/certified mail. The U. S. Postal Service operates a Post Office at Stennis to handle personal mail and postage.

6.3.1.3   Equipment Management

Maintains Government property records; controls tagging of Government property; disposes of Government property; administers warranty service, furniture pool; manages recycling program and the project's equipment acquisition and modernization programs. Provides disposal of government property, transfer of computer equipment to learning institutions and sale of government property through the General Services Administration (GSA).

6.3.2 Institutional Services

6.3.2.1   Food Services

Provides central cafeteria services for an a-la-carte breakfast and a full-service lunch menu five days a week; provides a site-maintenance satellite cafeteria, a mobile food service, and on-site catering services.

6.3.2.2   Transportation

Provides daily taxi-fleet operation, Visitors Center Tour Bus operation, special-request chauffeur support, airport vehicle dispatch operation, and GSA vehicle administration. Provides movement of furniture, equipment and personnel through a drayage crew.

6.3.2.3   Fire Department

Provides continuous 24-hr/day fire protection; conducts building, structure, and automatic system inspections; participates in facility design reviews; issues flame permits; provides standby coverage for operations as requested/required; provides CPR and fire prevention training; operates an ambulance service for SSC with certified Emergency Medical Technicians (EMTs) for medical emergencies.

6.3.2.4   Medical Clinic and Wellness/Fitness Facility

During daytime working hours, provides emergency and therapeutic, preventative medicine, employee assistance, industrial hygiene/environmental health, and wellness/fitness services to SSC personnel.

6.3.2.5   Custodial Services

Provides custodial service to SSC buildings and performs special cleaning services upon request.

6.3.2.6   Multimedia Services

Provides a comprehensive range of digital black & white and color graphics, publications, printing, and photographic services.

6.3.3 Information Services

6.3.3.1   Visitors Center

Operates the SSC Visitors Center. Conducts daily and special VIP tours and provides space-related programs.

6.3.3.2   Educator Resource Center

Supports the community's need for information and classroom instructional materials on math and science; develops workshops for area teachers; and provides teachers with video tapes and photo slides on space and science.

6.3.3.3   Conference Support

Provides conference support with audiovisual assistance in the four NASA conference rooms. Provides support to SSC off-site conferences, workshops, and lectures.

6.3.3.4   Media Support

Provides media coverage for the NASA/SSC Public Affairs Office and writes/edits/publishes the SSC newsletter, *Lagniappe*, each month.

6.3.4 Procurement

This function is performed by two specialized Branches within the Division - Purchasing and Contracts & Legal. Purchasing is responsible for obtaining supplies and services in support of 26,000 line items of inventory and the three prime NASA contractors. A sophisticated software system enhances the ability to transact over 18,000 orders annually and provides real-time interface with Finance, and Shipping and Receiving. Over the years, the vendor listing of qualified suppliers has grown to over 3,000, including the use of GSA.

The second Branch, Contracts and Legal, is responsible for those contracts of a more complex nature, such as those requiring a source evaluation. This typically includes service, cost reimbursement, and requirement-type contracts. In addition, the Legal Counsel is responsible for developing procurement policy in accordance with FAR and handling legal issues that arise as an adjunct of the procurement system.

The procurement system is kept current and is approved annually by the regional Defense Contract Audit Agency (DCAA). Without sacrificing price or delivery, Small and Small Disadvantaged Businesses are awarded approximately 65% and 15% of the dollar value of procurements, respectively, which enhances the stature of the system with local business and gives them confidence in our commitment to the local economy.

6.4   Technical Services

Technical Services include a Fluid Component Processing Facility (FCPF), Weld/Fabrication and Machine Shops, Component Support, Test Operations, Test Liason, and Advanced Programs.

6.4.1 Fluid Component Processing Facility

FCPF services all pressure and cryogenic components and utilizes an environmentally controlled clean room. This facility provides for the service, repair, cleaning, testing, and certification of mechanical components, systems, and parts in support of propulsion test activities. Hydrostatic testing can be performed up to 30,000 psi and pneumatic testing up to 15,000 psi. Cryogenic testing of components, up to 30 in. in diameter, can be performed down to $-320\ °F$ using LN_2 Assembly of components is performed inside a level-10K clean room.

6.4.2 Weld/Fabrication Shop

The Weld/Fabrication Shop provides gas welding or brazing and electric welding in accordance with ASME and American Welding Society (AWS) standards. Welders are certified to weld a wide range of metals (carbon steel, aluminum, and stainless steel), employing Shielded Metal Arc Welding (SMAW), Gas Tungsten Arc Welding (GTAW), stick welding, and brazing processes. The most up-to-date equipment [Computer Numerical Control (CNC) plasma arc cutter] is used. The shop can shear, roll, and press-form carbon steel up to ¾ in. and stainless steel up to ¾ in. Also, the shop is certified to make or repair ASME-coded pressure vessels. The shop's capabilities consist of fabrication of sheet metal, structural members, panels, template plates, HP piping, thin wall piping, vessels, and all types of experimental prototype components, assemblies, and systems. HP and cryogenic pipes are also fabricated and maintained.

6.4.3 Machine Shop

The Machine Shop provides services for fabrication, repair, and modifications involving plastics, nylon, and metals. The shop is capable of fabricating complete assemblies from blueprint or sketch to the finished product. The shop is equipped with two, Hurco, three-dimensional, computer-controlled machinery centers having a 3 ft capacity along each axis. Other shop equipment performs profile-lathe operations on 36 in. diameter by 22 ft length stock, and precision grinding. The shop has heat-treating capability for items up to 2 ft x 2 ft x 3 ft. A vertical mill provides milling tolerance to a 120 in. width and precision machining of parts to tolerances of ±0.0001 in. The shop operates lathes, milling machines, boring mills, shapers, drill presses, and grinders.

6.4.4 Component Support

Component Support provides professional engineering services of a functional nature in direct support of the test complex. The multifunctional disciplines assigned are divided into component documentation and application engineers, machining technology engineers versed in computer numerically controlled equipment, and welding engineers experienced in ASME and the American National Standards Institute (ANSI) code applications. These engineers provide the following functional services:

a. Prepare and maintain current procedures for maintenance cleaning and calibration of HP and cryogenic system components

b. Provide LOX and GOX expertise for materials compatibility

c. Perform system/component failure analysis for cryogenic system failures

d. Provide improved component/system designs and/or modifications resulting from failure analysis

e. Provide spares provisioning for systems of responsibility

f. Provide technical assistance to shops/associate contractors as required

g. Provide for configuration management and control coordination with Engineering Services Division

6.4.5 Test Operations

Test Operations consists of project management and design engineers who provide rapid response engineering support to the Test Complex. This group prepares designs, construction specifications, and cost estimates for facility modifications within the Test Complex and other propulsion testing programs. Using the Area Engineer Concept, the group identifies requirements, initiates necessary work, and follows the implementation of such work to completion. This group also oversees the Pressure Vessel Recertification Program for SSC including ASME Coded Pressure System designs and repairs. Engineers, supporting Maintenance Engineering, oversee the maintenance of facilities, systems, and equipment, and conduct annual facility inspections to ensure that needed maintenance is identified.

6.4.6 Test Liaison

Test liaison coordinates implementation of repair maintenance and special projects within the Test Complex using shop forces. Liaison provides direction, planning, estimating, and oversight of all work including construction reviews and technical recommendations. All work is closely coordinated with other contractors on site.

6.4.7 Advanced Programs Group

The Advanced Programs Group provides support to the Propulsion Test Directorate and NASA Technology Transfer Office. It oversees the operation of the Seal Configuration Tester to test and evaluate various softgoods and seals in various pressurized systems.

6.5 Nondestructive Test and Evaluation Laboratory [22]

The NDTE Laboratory furnishes support in all aspects of nondestructive testing and evaluation of structural and material integrity investigations. These include material inspections for surface and internal flaws, identification and quantification of gas leakage from pressurized gas systems and the verification of "fitness-for-service" of facility systems. All SSC pressure vessels and barges are periodically inspected and re-certified as mandated by applicable codes and standards. Periodically, the NDT/E Laboratory is requested to conduct nondestructive testing and evaluation of SSME components.

This laboratory is capable of performing inspections and diagnostics both in the field and in the laboratory. To accomplish its many tasks, the laboratory has an extensive inventory of test equipment and certified technicians for performing a variety of NDT/E methods. Some of the laboratory's inspection and/or diagnostic capabilities include

112

radiography, ultrasonic, liquid penetrant, and visual examination including boroscopy, magnetic particle, mass spectrometer leak detection and eddy-current testing.

Radiography as well the other NDTE methods in use at SSC has been performed on various SSME components, including the main LOX inlet manifold seam welds, high pressure gas storage vessels and pressure piping, cryogenic storage vessels and cryogenic transfer lines. When required, mobile film-processing facilities are available to support field radiographic operations. Equipment in service at SSC NDT/E Laboratory includes but is not limited to the following as tabulated in Table 6.5-1.

| NDTE Major Equipment Listing ||
Equipment	Function
Boroscopic inspection system	Inspects interior of vessels, piping, etc.
Leak detector (portable)	Helium mass-spectrometer leak detection system
Ultrasonic Flaw Detection	Detects flaws in material and welds
Pancake GM detector	Chamber for radioactive leak wipe analysis
Liquid penetrant testing system	Testing of metals for surface defects
Magnetic particle testing coils	Testing of ferromagnetic metals (circular)
Magnetic particle testing yokes	Testing of ferromagnetic metals (longitudinal)
New Age Versitron hardness tester	Hardness testing of materials (Rockwell C)
Radiographic darkroom trailers	Film processing
Radiographic exposure device (cobalt-60)	Examines metals up to 9 in. thick
Radiographic exposure device (iridium 192)	Examines metals up to 3.5 in. thick
Smart eddy 2.0	Eddy current testing
Spectrometer (Ludlum 2600)	Single channel analyzer for Pancake GM detector
Temperature testing probe	Tests temperatures of materials
Thiokol Equotip hardness tester	Hardness testing of materials (portable)
Troxler moisture density gauge	Measures densities in soils, asphalt, concrete, etc.
Ultrasonic thickness gauges	Measures material thickness
X-ray imaging system [real time (100 kV)]	Fluoroscopic x-ray examination of metals
X-ray system [Andrex (160 kV)]	X-ray of materials up to 1 in. thick (portable)
X-ray system [inner view (60 kV)]	Detailed inspection of electronic components
Liquid penetrant system (Zyglo)	Surface inspection of small parts

Table 6.5-1 NDTE Laboratory Equipment

6.5.1 Radiographic Inspection (RT)

A radiograph is basically a two-dimensional picture of the intensity distribution of some form of radiation projected from a source that has passed through a material object that partially attenuates the intensity of radiation. Voids, changes in thickness, or regions of different composition will attenuate the radiation by different amounts, thus producing a projected shadow of themselves. Using x-rays and gamma rays for the penetrating radiation, the radiographic information is captured on film or fluorescent screens. In real-time systems, a television camera collects the image on a fluorescent screen and the image is subsequently digitized for enhancement analysis and storage.

6.5.2 Ultrasonic Testing (UT)

Ultrasonic testing is a nondestructive method wherein beams of ultra high frequency sound waves are introduced into a test object to detect and locate surface and internal discontinuities. Detection, location, and evaluation of discontinuities are possible as the velocity of sound through a given material is nearly constant. This makes distance measurements possible. The amplitude of the reflected pulse is nearly proportional to the size of the reflector. Material thickness measurement is accomplished by measuring the time required for the pulse to travel through the material, reflect off the opposite surface, and return.

6.5.3 Liquid Penetrant Testing (PT)

Liquid penetrant inspection is a process for locating flaws that are open to the surface in solid, essentially nonporous materials. The test article in the area of interest is first coated with the penetrant solution. After a certain dwell time, the excess penetrant is removed. Any flaws open to the surface will retain a small portion of the penetrant. A developer is then applied to the test surface. Due to capillary action, the remaining dye will be drawn to the surface of the object being inspected. The results are then interpreted. Except for visual inspection, liquid penetrate is perhaps the most commonly used nondestructive test for examination of nonmagnetic parts.

6.5.4 Magnetic Particle Testing (MT)

Nondestructive magnetic particle testing is used to detect surface or near-surface discontinuities in magnetic materials. The method is based on the principle that the magnetic lines of force in a ferromagnetic material will be distorted by the change in material continuity. If a discontinuity exists in a magnetized material, surface or otherwise, the magnetic force lines will be distorted. When fine magnetic particles are spread over the area of discontinuity, they will accumulate at the discontinuity. This accumulation of particles will be visible under proper lighting conditions.

6.5.5 Leak Testing (MSLD)

Mass spectrometer leak detection is a method utilized to ensure pressure system or vacuum integrity. This is an extremely sensitive test method in which helium is used as a tracer gas. In pressure systems, the test object is pressurized using a gas mixture including a minimum of 10% helium within the mixture. The test object is then scanned with the MSLD unit and any leakage noted. When testing a vacuum system, the test article is evacuated with the MSLD unit sampling the vacuum annulus. Helium is then introduced as a blanket around the article being tested and any leakage into the test object will be observed.

6.5.6 Eddy-Current Testing (ET)

Eddy-current testing is a method used to locate surface or subsurface flaws in electrically conductive materials and to evaluate material characteristics; for example, hardness, heat-treat, other metallurgical conditions. The test article is brought into a time-varying electromagnetic field that induces electrical current into the test article. The force field of the electric current resembles eddies that are seen in the turbulent waters of a flowing stream. The amount of electrical current passing through a test article is determined by the electrical conductivity of the test article, as well as the frequency and amplitude of the magnetic field. Eddy-current testing is used to detect surface and subsurface flaws, irregularities in material structure, and variation in chemical composition in metallurgy.

6.5.7 Visual Inspection (VT)

Visual inspection is a nondestructive method of evaluating materials for the detection of discontinuities that are open to the surface; for example, cracks, seams, laps, cold sheets, laminations and porosity. With the advent of Digital Photography, an extensive database of photos documenting various system conditions is being assembled. Remote visual inspection is possible utilizing a boroscope or similar equipment. Visual inspection is the initial method used in evaluating the condition of cryogenic storage and pressure vessels, pipelines and valves, liquid propellant transfer barges, and test engines. The three methods of visual inspection by the NDTE Laboratory include direct, remote, and translucent.

6.5.8 Hardness Testing (VT)

Hardness testing is a mechanical test for evaluating the properties of metals and certain other materials. Hardness testing is used at SSC to determine if purchased and/or contractor construction materials meet the design specifications. Such materials can be piping, beams, vessels, and plating. Hardness testing is also used in the analysis of cause when determining a system failure, if such testing is applicable. The major hardness tests used are indentation, scratches, cutting, abrasion, and erosion.

6.5.9 Bubble Leak Testing (LT)

Leak detection is a direct pressure-testing method used to locate leaks in a pressurized component by the application of a solution that will form bubbles as leaking gas passes through the component. Leak detection testing at SSC is used to detect leaks in the site-wide piping systems, pressure vessels, vacuum jacketed lines, storage vessels, and the propellant-transfer barges.

7.0 OUTSOURSING DESKTOP INITIATIVE FOR NASA (ODIN)

The ODIN Contractor provides computer desktop, data communications, telecommunications, administrative radio, onsite television network circuits, and other outsourcing support services to all Stennis agencies.

7.1 Office Automation

The Office Automation Software that supports the test facility is referred to as Stennis Desktop Services (SDS). SDS is built around the Microsoft Office product Office Suite and is implemented as the NASA SSC standard for office automation. These services allow for e-mail, scheduling, calendar management, and access to the Internet. Included in the SDS are standard PC applications to assist the individual. These applications include Microsoft Word, Microsoft Excel, Microsoft PowerPoint and Microsoft Project. All the above services operate in a Microsoft Windows environment. The host of the SDS is an ODIN Desktop Seat. These vary in capability and include PCs, MACs, and Unix based platforms.

7.2 Telecommunications

ODIN provides state of the art telecommunications services to SSC, including:

a. Telephone (digital & analog) and Voice Messaging
b. Switched and dedicated data circuits
c. LAN network connectivity, off-site access
d. Trunked land-mobile radio service with telephone connectivity
e. Site-wide CATV video distribution with satellite down-links from major information services (CNN, Headline News, etc)
f. Video Teleconferencing

Telecommunications services, other than radio, are provided through a mix of copper and fiber cable plant. Most new services are routed on a combination of single-mode and multi-mode fiber plant supporting bandwidths from 1 mbs to 2.1 gbs. Current reliability measurements of the telephone system are 99.999% reliability.

7.2.1 Telephone System

The telephone system provides digital & analog, dedicated & switched voice and data service as well as low speed data connectivity up to 56 kbs. The system was recently enhanced to support Primary Rate Interface ISDN (Integrated Services Digital Network) in late 1995. The system currently supports approximately 5200 user ports with the capability of supporting 16,000 ports.

Special high-speed dedicated data circuits are provided at speeds ranging from 56 kbs to 45 mbs. Engineering, design, and implementation of all levels of data services are supported as well as operational maintenance.

7.2.2 LAN

ODIN also provides and supports a switched LAN environment of approximately 2500 computers belonging to NASA, NASA contractors, and resident agencies. Engineering, design, and implementation of all network requirements are done within the Telecommunications group. Protocols that are currently approved for use on the SSC LANs are:

a. TCP/IP
b. UDP/IP
c. AppleTalk

7.2.3 Circuits and Wiring

ODIN is also responsible for the engineering, design, implementation, and maintenance of all premise distribution systems throughout SSC. Currently, category 5 unshielded twisted pair, capable of 100 Mbps transfer rates, is being used as the standard for wiring to the end user environment. The distribution system is further supported by intelligent networking hubs capable of providing a minimum of switched 10 Mbps to the desktop, remote monitoring (SNMP) to the node level, and problem resolution via internal diagnostic capabilities. These switches are routed back to the Telecommunications complex over the site wide fiber distribution system utilizing ATM (Asynchronous Transfer Mode) standard LANE 2.0 at 155 Mbps and 622 Mbps transfer rates.

Off-site WAN access is provided through a series of firewall and/or routers to the NASA Integrated Systems Network (NISN). The NISN provides such services as Internet access as well as primary network support for Space Shuttle Main Engine (SSME) test data. ODIN is not responsible for working these off site requirements; however, they are responsible for on-site coordination.

7.2.4 Network Operations Center

ODIN also operates a Network Operations Center (NOC), which proactively monitors the health of SSC network systems and gathers appropriate usage statistics for analysis. This center performs all first and second level maintenance on all communications systems. The maintenance crew is backed by third level vendor maintenance contracts and support as required. All maintenance activities, including preventive and corrective maintenance, are coordinated to maximize system availability and minimize user impacts. Extended coverage, 6:00am - 6:00pm is provided for Telecommunication systems.

7.2.5 Wireless Communications

ODIN also operates the SSC Motorola Trunked UHF radio system supporting approximately 500 fixed, mobile, and portable radios in approximately 40 active radio networks. Off site telephone connectivity is provided to some radio networks (e.g. Fire & Security, Disaster Recovery, Emergency Operations Center, etc.), to coordinate activities with off site emergency service providers.

Pagers are not provided under the ODIN contract but are provided by NASA and administered through TTSC support services.

7.2.6 Video Network

Current news and weather information and educational television programming are provided through the SSC Video Network. This system uses standard CATV and studio components to provide 17 channels of video service to approximately 85 SSC locations. Programming consists of fixed information services (e.g. CNN, The Weather Channel, NWS NEXRAD radar images), received from off site in addition to local information such as local weather gauges and the SSC weather radar system. Other services (commercial television broadcasts), are provided on a demand basis. In addition, the system is configured to allow remote origination of video programming (e.g. engine test firings, Public Affairs programming etc.) from end user locations back to the telecommunications complex for processing and retransmission across SSC. Satellite downlink capabilities are provided within the ODIN contract also.

Low Bandwidth Video (LBV), is also offered for as an inexpensive means for periodic video conferencing requirements. Portable units are available and can be located anywhere on SSC that contains premise distribution wiring. The LBV is not provided under the ODIN contract but is provided by NASA.

7.2.7 Desktop Computers

Another valuable component of ODIN is the provision and maintenance of ADPE (Automated Data Processing Equipment). The ADPE unit provides desktop to desktop repair of PCs, servers, workstations, printers and various other peripheral equipment as well as installation and maintenance of all NASA standard software suites supporting office automation. Currently the group supports approximately 6700 pieces of equipment throughout SSC which includes institutional and demand agencies, i.e. NASA, NASA Contractors, SSME, etc.

7.2.8 Help Desk

ODIN is also responsible for the help desk for IT support. Troubles or status information on requests can be obtained by calling the on-site help desk. Responses to troubles are a function of the service level being procured for a user's IT (phone, desktop, etc).

8.0 OTHER

8.1 Cray Supercomputer [26]

The Cray Y-MP8/8128 supercomputer with a UNIX-based operating system is the Navy's most powerful computing system, and is available as an SSC resource for computer modeling/analysis. It has eight processors, which working individually or collectively on a computation share a 128 M, 64 bit, main word memory. This memory capability amounts to over one billion alphanumeric characters. At peak operating speed, the system is capable of achieving 2.67 billion floating point operations per second.

The Cray supercomputer has the capability of meeting the large-volume computational requirements of advanced propulsion testing and development.

APPENDIX A

REFERENCE DOCUMENTS AND DRAWINGS

REFERENCE DOCUMENTS

[1]   *Deleted*

[2]   *Deleted*

[3]   *Component Test Facility Design Basis Document* for John C. Stennis Space Center, Bechtel National, Inc., Oak Ridge, TN, March 10, 1989.

[4]   *Consolidation Report for the Storage Vessel Requirements Studies and Reports and Cryogenic Support Systems*, Report 910-89-051, Pan Am World Services, Inc., John C. Stennis Space Center, Stennis Space Center, MS 39529, September 12, 1989.

[5]   *Facilities Master Plan*, National Space Technology Laboratories, NSTL Station, MS 39529, May 1979.

[6]   *Deleted*

[7]   "High Pressure Industrial Water and Emergency Electric Power Plant Capabilities Statement," Sverdrup Technology, Inc., Stennis Space Center, MS 39529, September 11, 1991.

[8]   *Deleted*

[9]   *John C. Stennis Space Center Information Brochure*, John C. Stennis Space Center, Bay St. Louis, MS 39529, 1990.

[10] *Deleted*

[11] *Deleted*

[12] *Meteorological Forecasting Plan—Advanced Solid Rocket Motor Program*, John C. Stennis Space Center, Stennis Space Center, MS 39529, December 18, 1990.

[13] "Mississippi Test Facility, Package B, Final Submittal, Volume I," Sverdrup & Parcel and Associates, Inc., Engineers-Architects, August 26, 1963.

[14] *Deleted*

[15] "Mississippi Test Facility S-IC Complex Briefing Data" for George C. Marshall Space Flight Center, Huntsville, Alabama, April 1965. Report 937-64-002.

[16] "Mississippi Test Facility S-II Complex Briefing Data" for George C. Marshall Space Flight Center, Huntsville, Alabama, April 1965. Report 937-64-004.

[17]   *Deleted*

[18]   *Deleted*

[19]   *Deleted*

[20]   *Deleted*

[21]   *Deleted*

[22]   *Non-Destructive Testing at John C. Stennis Space Center*, Sverdrup Technology, Inc., Stennis Space Center, MS 39529, Charles Schimmel, Jr., P.E. and E.J. Casanova, July 30, 1991.

[23]   *Deleted*

[24]   *NSTL Sitewide High Pressure Gas Distribution Systems*, Pan Am World Services, National Space Technology Laboratories, NSTL, MS 39529, March 26, 1985. Report 911-85-001.

[25]   *Deleted*

[26]   *Primary Oceanographic Prediction System (POPS) CRAY Y-MP8/8128 Large Scale Computer (LSC)*, Naval Oceanographic Office, Super Computer Center Division, Code-AM, Stennis Space Center, MS 39529, March 23, 1992.

[27]   *Propellant, Pressurants, Storage Facilities, Etc.*, Sverdrup Technology, Inc., Stennis Space Center, MS 39529, F. Gasporovic, August 14, 1990.

[28]   "Propulsion Test Operations Test Capability Document," John C. Stennis Space Center, Bay St. Louis, MS 39529.

[29]   *Deleted*

[30]   "Reliability Analysis, NSTL Test Complex Electrical Distribution, 13.9 kV 480 V Breakers," Sverdrup Technology, Report 937-87-010, September 1987.

[31]   Response to the *Roles and Missions* Report, Administrator's Decision Memorandum, "NASA Roles and Missions," NASA Office of the Administrator, Washington, D.C. 20548, December 30, 1991.

[32]   "Rocketdyne Test Control Center Data Recording Capabilities," John C. Stennis Space Center, January 17, 1992.

[33]   S-IC Stage Test Program, Mississippi Test Facility, Bay St. Louis, MS, Test & Quality Evaluation Office, October 29, 1970.

[34] S-IC Test Complex, Test Position B-2, Construction Package "F", Design Analysis, Vol. IV—Section C. Sverdrup & Parcel and Associates, Inc., Engineers—Architects, 1964.

[35] *S-II Stages Static Firing Summary*, North American Rockwell Corporation, Space Division, Mississippi Test Operations.

[36] *Deleted*

[37] *SSC Pressure Vessel and Systems Recertification Plan*, NASA/SSC Operating Procedure 83-1, July 15, 1982, Revision 4.

[38] *NSTL Sitewide High Pressure Gas Distribution Systems*, SSC Report Number 911-85-001, Pan Am World Services, Inc., National Space Technology Laboratories, NSTL, MS 39529, March 26, 1985.

[39] *Deleted*

[40] "Static Testing of Apollo/Saturn V First Stage (S-IC) at Mississippi Test Facility," Background Brief, Public Affairs Office, MTF, Bay St. Louis, MS.

[41] "Static Testing of Apollo/Saturn V Second Stage (S-II) at Mississippi Test Facility," Background Brief, Public Affairs Office, MTF, Bay St. Louis, MS.

[42] *Stennis Space Center High Pressure Gas Facility Operations Automation Requirements Study*, Stennis Space Center, Stennis Space Center, Ms 39529, Dale Larson, Report 4525-92-002, May 5, 1992.

[43] *Study of the SSC Gas and Propellant Systems* for NASA Facilities Engineering Division, Pan Am World services, Inc., John C. Stennis Space Center, Stennis Space Center, MS 39529, Report 910-88-047, January 20, 1989.

[44] "Propulsion Test Directorate E1 Facility Capabilities Document", E1-FCD-001, Revision 1, John C. Stennis Space Center, MS 39529, May, 2001.

[45] "Propulsion Test Directorate E2 Cell 1 Facility Capabilities Document", E2-FCD-001, Revision 1, John C. Stennis Space Center, MS 39529, Feb. 2001.

[46] "Propulsion Test Directorate E2 Cell 2 Facility Capabilities Document", E2-FCD-002, Revision 0, John C. Stennis Space Center, MS 39529, May 2001.

[47] "Propulsion Test Directorate E3 Facility Capabilities Document", E3-FCD-001, Revision 1, John C. Stennis Space Center, MS 39529, June, 2001.

REFERENCE DRAWINGS

(1)   11BGK-P001—Stennis CTF Project—Piping and Instrument Diagram, LO Unloading, Storage, and Transfer.

(2)   11BGK-P002—Stennis CTF Project—Piping and Instrument Diagram, LO, LP, and HP Run Tanks, Sheet 1 of 2.

(3)   11BGK-P002—Stennis CTF Project—Piping and Instrument Diagram, LO, LP, and HP Run Tanks, Sheet 2 of 2.

(4)   11BGK-P003—Stennis CTF Project—Piping and Instrument Diagram, LO Turbopump Assembly Test Cell, Sheet 1 of 3.

(5)   11BGK-P003—Stennis CTF Project—Piping and Instrument Diagram, LO Turbopump Assembly Test Cell, Sheet 2 of 3.

(6)   11BGK-P003—Stennis CTF Project—Piping and Instrument Diagram, LO Turbopump Assembly Test Cell, Sheet 3 of 3.

(7)   11CDK-P001—Hydrogen Liquid—HPGF Hydrogen Shelter Piping, Schematic, Sheet 1 of 3.

(8)   11CDK-P001—Hydrogen Liquid—HPGF Hydrogen Shelter, Air Products Storage System, Utility Flow Diagram, Sheet 2 of 3.

(9)   11CDK-P001—Hydrogen Liquid—HPGF Hydrogen Shelter, Utility Flow Diagram, Sheet 3 of 3.

(10)   11CGD-P001—Hydrogen Gaseous—A-1 Pressure-Reducing Area Piping Schematic, Sheet 1 of 1.

(11)   11CGD-P002—Hydrogen Gaseous—S-II A-1 Test Stand, Main Supply Piping Schematic, Sheet 1 of 1.

(12)   11CGD-P003—Hydrogen Liquid & Gas—S-II Test Complex LH2 Barge Dock, Test Stand and Barge Flare Stack Piping Schematic, Sheet 1 of 1.

(13)   11CGD-P004—Hydrogen Liquid—S-II A-1 Test Stand, Main Supply Piping Schematic, Sheet 1 of 1.

(14)   11CGF-P001—Hydrogen Gaseous—A-2 Test Stand, Pressure-Reducing Area Piping Schematic, Sheet 1 of 1.

(15)   11CGF-P001—Hydrogen Gaseous—A-2 Test Stand, Pressure-Reducing Area Piping Schematic.

A-7

(16) 11CGF-P002—Hydrogen Gaseous—S-II A-2 Test Stand, Main Supply Piping Schematic, Sheet 1 of 1.

(17) 11CGF-P003—Hydrogen Liquid—S-II A-2 Test Stand, LH2 Dock and Flare Stack Piping Schematic, Sheet 1 of 1.

(18) 11CGK-P001—Stennis CTF Project—Piping and Instrument Diagram, LH2 Trailer Unloading, Storage, and Transfer, Sheet 1 of 2.

(19) 11CGK-P001—Stennis CTF Project—Piping and Instrument Diagram, LH2 Trailer Unloading, Storage, and Transfer, Sheet 2 of 2.

(20) 11CGK-P002—Stennis CTF Project—Piping and Instrument Diagram, LH2 LP and HP Run Tanks, Sheet 1 of 2.

(21) 11CGK-P002—Stennis CTF Project—Piping and Instrument Diagram, LH2 LP and HP Run Tanks, Sheet 2 of 2.

(22) 11CGK-P003—Stennis CTF Project—Piping and Instrument Diagram, LH2 Turbopump Assembly Test Cell, Sheet 1 of 3.

(23) 11CGK-P003—Stennis CTF Project—Piping and Instrument Diagram, LH2 Turbopump Assembly Test Cell, Sheet 2 of 3.

(24) 11CGK-P003—Stennis CTF Project—Piping and Instrument Diagram, LH2 Turbopump Assembly Test Cell, Sheet 3 of 3.

(25) 11CGK-P004—Stennis CTF Project—Piping and Instrument Diagram, HP and UHP GH2 Storage, Sheet 1 of 1.

(26) 11CGK-P006—Stennis CTF Project—Piping and Instrument Diagram Flare Stacks, Sheet 1 of 1.

(27) 11CG0-P001—Hydrogen—S-II Test Complex, Hydrogen Gas Battery Area Piping Schematic, Sheet 1.

(28) 11FDK-P001—Nitrogen System—HPGF Compressor Building, Nitrogen Pump Area Piping Schematic, Sheet 1 of 4.

(29) 11FDK-P001—Nitrogen System—HPGF Compressor Building, Nitrogen Pump Area Piping Schematic, Sheet 2 of 4.

(30) 11FDK-P001—Nitrogen System—HPGF Compressor Building, Nitrogen Pump Area Piping Schematic, Sheet 3 of 4.

(31) 11FDK-P001—Nitrogen System—HPGF Compressor Building, Nitrogen Pump Area Piping Schematic, Sheet 4 of 4.

(32) 11FGD-P001—Nitrogen—S-II A-1 Test Stand, Pressure-Reducing Area Piping Schematic, Sheet 1 of 1.

(33) 11FGD-P002—Nitrogen—S-II A-1 Test Stand, Main Supply Piping Schematic, Sheet 1 of 1.

(34) 11FGF-P001—Nitrogen—S-II A-2 Test Stand, Pressure-Reducing Area Piping Schematic, Sheet 1 of 1.

(35) 11FGF-P003—Nitrogen—S-II A-2 Test Stand, Main Supply, Sheet 1 of 1.

(36) 11FGK-P001—Stennis CTF Project—Piping and Instrument Diagram, LN2 Trailer and HP Pump Vaporization, Sheet 1.

(37) 11FGK-P001—Stennis CTF Project—Piping and Instrument Diagram, HP and UHP GN2 Storage.

(38) 11FGK-P002—Stennis CTF Project—Piping and Instrument Diagram, GN2 Distribution System, Sheet 1, Sheet 2.

(39) 11FG0-P001 (Rev. 6)—Nitrogen—Complex A, Inert Gas Battery Area Piping Schematic, Sheet 1 of 1.

(40) 11GDK-P001 (Rev. 8)—Helium Gaseous—HPGF Compressor Loading and Storage Flow Diagram, Sheet 1 of 5.

(41) 11GDK-P001—Helium Gaseous—HPGF Compressor Building, Purification, Sampling, and Supply, Sheet 5 of 5.

(42) 11GGD-P001—Helium—S-II A-1 Test Stand, Inert Gas Pressure-Reducing Area Piping Schematic, Sheet 1 of 1.

(43) 11GGD-P002—Helium—S-II A-1 Test Stand, Main Supply Piping Schematic, Sheet 1 of 1.

(44) 11GGF-P001—Helium—S-II A-2 Test Stand, Inert Gas Pressure-Reducing Piping Schematic, Sheet 1 of 1.

(45) 11GG0-P001—Helium—Test Complex "A," Inert Gas Battery Area Piping Schematic, Sheet 1 of 1.

(46) 11HDK-P001—High Pressure Air System—High Pressure Gas Facility Compressor Building Piping Schematic, Sheet 1 of 1.

(47) 11HG0-P001—High Pressure Air—S-II Test Complex, Inert Gas Battery Area Piping Schematic, Sheet 1 of 1.

(48) 11HGD-P001—High Pressure Air—S-II A-1 Test Stand, Main Supply Piping Schematic, Sheet 1 of 1.

(49) 11HGF-P001—High Pressure Air—S-II A-2 Test Stand, Inert Gas Pressure-Reducing Area Piping Schematic.

(50) 11HGF-P002—High Pressure Air—S-II A-2 Test Stand, Main Supply Piping Schematic, Sheet 1 of 1.

(51) 12B00-E001—SSC Electrical Distribution System—13.9-kV Single Line Diagram.

(52) 31B00-K001—LOX Barges 1-6 Block Diagram—Advanced Schematic, Sheet 1.

(53) 31B00-P001—Marine Equipment LOX Barge No. 1—LOX Piping Schematic (Typical) P/L Issued.

(54) 31C00-P001—Marine Equipment LH2 Barge No. 1—Hydrogen Piping Schematic (Typical).

(55) *Deleted*

(56) 110GK-K006—Stennis CTF Project—Single Line Diagram.

(57) 110GK-R004—Stennis CTF Project—Test Stand Sections, 4 Sheets.

(58) 200GK-A308—Stennis CTF Project—Test Cell/SCB Foundation Plan Sections, 1 Sheet.

(59) 460GK-H011—Stennis CTF Project—Site Plan.

(60) *Deleted*

(61) 11CGD-P003, P004—Modification to A-1 Test Stand—Liquid Hydrogen System Flow Diagram.

(62) 11CGD-P003, 11FGD-P003—Modification to A-1 Test Stand—Liquid Hydrogen System Plan.

(63) 11FG0-P001, 11FGF-P001—S-II Test Complex, Test Stand A-1—High-Pressure Nitrogen Flow Diagram.

(64)   11FGD-P001, P003—S-II Test Complex, Test Stand A-1—High-Pressure Nitrogen Flow Diagram.

(65)   11FGD-P002, P004—S-II Test Complex, Test Stand A-1—High-Pressure Nitrogen Flow Diagram.

(66)   11GG0, P001—S-II Test Complex, Test Stand A-1—High-Pressure Helium System Flow Diagram.

(67)   11GGD-P001,11CGD-P002—S-II Test Complex, Test Stand A-1—High-Pressure Helium System Flow Diagram.

(68)   11CG0-P001, 11CGF-P001—S-II Test Complex, Test Stand A-1—High-Pressure Hydrogen System Flow Diagram.

(69)   11CGD-P001, 11CGD-P002—S-II Test Complex, Test Stand A-1—High-Pressure Hydrogen System Flow Diagram.

(70)   11HG0-P001, 11CGF-P001—S-II Test Complex, Test Stand A-1—High-Pressure Air System Flow Diagram.

(71)   11HGD-P001, 11HGD-P002—S-II Test Complex, Test Stand A-1—High-Pressure Air System Flow Diagram.

(72)   11BGD-P001—S-II Test Complex, Test Stand A-1—Liquid Oxygen System Flow Diagram.

(73)   11CGD-P003, 11CGD-P004—S-II Test Complex, Test Stand A-1—Liquid Hydrogen System Flow Diagram.

(74)   11FG0-P001—S-II Test Complex, Test Stands A-1 and A-2—High-Pressure Nitrogen Flow Diagram.

(75)   11FGF-P001—S-II Test Complex, Test Stand A-2—High-Pressure Nitrogen Flow Diagram.

(76)   11CG0-P001—S-II Test Complex, Test Stands A-1 and A-2—High-Pressure Helium System Flow Diagram.

(77)   11GGF-P001—S-II Test Complex, Test Stand A-2—High-Pressure Helium System Flow Diagram.

(78)   11GG0-P001—S-II Test Complex, Test Stands A-1 and A-2—High-Pressure Hydrogen System Flow Diagram.

(79) 11CGF-P001—S-II Test Complex, Test Stand A-2—High-Pressure Hydrogen System Flow Diagram.

(80) 11HG0-P001—S-II Test Complex, Test Stands A-1 and A-2—High-Pressure Air System Flow Diagram.

(81) 11HGF-P001—S-II Test Complex, Test Stand A-2—High-Pressure Air System Flow Diagram.

(82) 11CGF-P001—S-II Test Complex, Test Stand A-2—Liquid Hydrogen System Flow Diagram.

(83) 11CGF-P003—S-II Test Complex, Test Stands A-1 and A-2—Flare System Flow Diagram.

(84) *Deleted*

(85) 200GF-A019—S-II Test Complex, Test Stand A-2—Test Stand General Arrangement.

(86) 11HB0-P001, 11FH0-P002, 11BHF-P001—S-IC Test Complex, Test Position B-2—High-Pressure Nitrogen System Flow Diagram.

(87) *Cancelled*

(88) *Cancelled*

(89) 11FHD-P001—S-IC Test Complex, Test Position B-1—Overall Flow Diagram—HP Gas Systems—Nitrogen.

(90) 11FHD-P001—S-IC Test Complex, Test Position B-1—Overall Flow Diagram—HP Gas Systems—Nitrogen System.

(91) 11HHD-P001—S-IC Test Complex, Test Position B-1—Overall Flow Diagram—HP Gas Systems—HP Air System.

(92) 11GHD-P001—S-IC Test Complex, Test Position B-1—Overall Flow Diagrams—HP Gas Systems—Helium.

(93) 11BHD-P001—S-IC Test Complex, Test Position B-1—Overall Flow Diagrams Cryogenic Systems—Liquid Oxygen.

(94) *Cancelled*

(95) 200HD-A001—S-IC Test Complex, Test Position B-1, Test Stand General Arrangement.

(96) MTF-SSE-G108—Modifications of Test Facilities for Space Shuttle Engine Testing—Test Stand A-2 General Notes.

(97) MTF-SSE-S14—Modifications of Test Facilities for Space Shuttle Engine Testing Thrust Measuring System—General Arrangement Test Stand A-1.

(98) MTF-SSE-S117—Modifications of Test Facilities for Space Shuttle Engine Testing Thrust Measurement System—General Arrangement Test Stand A-2.

(99) 11CGF-P001—Modifications for Orbiter Propulsion System Test Facilities, Phase II—General Notes.

(100) EMI FUB P0100—S-II Test Complex, Test Stand A-2—Liquid Hydrogen System Flow Diagram.

(101) XDY90101 Figure A—Restoration of HPGF Existing Facility Site Plan, Building 3304 and 3305.

(102) EMI H9A0D00020—High Heat Flux Rig Site Plan.

APPENDIX B

ACRONYMS AND ABBREVIATIONS

ACRONYMS

—A—

ALS	Advanced Launch System
ANSI	American National Standards Institute
ASME	American Society of Mechanical Engineers
ASRM	Advanced Solid Rocket Motor
A-TCC	"A" Test Control Center
ATM	Asynchronous Transfer Mode
AWS	American Welding Society

—B—

B-TCC	"B" Test Control Center

—C—

CADD	Computer Aided Design and Drafting
CCTV	Closed Circuit Television
CD	Compact Disk
CEF	Central Engineering File
CNC	Computer Numerical Control
C of F	Construction of Facilities
CS	Control System
CTF	Component Test Facility
CW	Cooling Water

—D—

DAF	Data Acquisition Facility
DAS	Data Acquisition System
DAT	Data Tape Format
DC	Direct Current
DI	Deionized
DCAA	Defense Contract Audit Agency
DO	Dissolved Oxygen
DVOM	Digital Voltage Ohm Meter

—E—

EEB	Electrical Equipment Building
EPA	Environmental Protection Agency

—F—

FAR	Federal Acquisition Regulations
FCPF	Fluid Component Processing Facility
FFT	Fast Fourier Transform
FOS	Facility Operating Services
FMS	Flow Measurement System
FRS	Financial Reporting System
FTIR	Fourier Transform Infrared

—G—

GH_2	Gaseous Hydrogen
GHe	Gaseous Helium
GC	Gas Chromatograph
GMAL	Gas and Material Analysis Laboratory
GN_2	Gaseous Nitrogen
GOX	Gaseous Hydrogen
GPM	Gallons per Minute
GSA	General Services Administration
GTAW	Gas Tungsten Arc Welding

—H—

HF	High Frequency
HHFF	High Heat Flux Facility
H_2O	Water
H_2O_2	Hydrogen Peroxide
HP	High-Pressure (3,000 psi to 8,000 psi)
HPA	High-Pressure Air
HPGF	High-Pressure Gas Facility
HPIW	High-Pressure Industrial Water
HSDAS	High Speed Data Acquisition System
HTDP	Hybrid Technology Demonstration Program
HVAC	Heating, Ventilating, and Air Conditioning

—I—

I/O	Input/Output
IRIG	Time standard accurate to 1 ms

—J—K—L—

JP	Jet Propellant
LAN	Local Area Network
LH$_2$	Liquid Hydrogen
LN$_2$	Liquid Nitrogen
LOX	Liquid Oxygen
LP	Low Pressure
LSDAS	Low Speed Data Acquisition System

—M—

MASS	Management Accounting Statusing System
MAWP	Maximum Allowable Working Pressure
MCC	Motor Control Center
MDWP	Maximum Design Working Pressure
MP	Medium Pressure (1,000 psi to 3,000 psi)
MPTA	Main Propulsion Test Article
MV	Motor Valve

—N—

NASA	National Aeronautics and Space Administration
NASP	National Aero-Space Plane
NDTE	Nondestructive Testing and Evaluation
NIST	National Institute of Standards and Technology
NLS	National Launch System
NVR	Nonvolatile Residue

—O—P—

ODIN	Outsourcing Desktop Initiative for NASA
PC	Personal Computer
PER	Preliminary Engineering Report
PLC	Programmable Logic Controller
PM	Preventive Maintenance
PM/CM	Preventive and/or Corrective Maintenance
PRA	Pressure Reducing Area
PSCS	Program Support Computer System

—Q—R—

RF	Radio Frequency
ROM	Read-Only Memory
RP	Rocket Propellant
RTD	Resistance Temperature Detector

—S—

SCB	Signal-Conditioning Building
SCFM	Standard Cubic Foot per Minute
SDS	Stennis Desktop Services
SEM	Scanning Electron Microscope
SSC	John C. Stennis Space Center
SSME	Space Shuttle Main Engine
STD	Standard
STE	Special Test Equipment
STS	Space Transportation System
SVHS	Super VHS

—T—

TCC	Test Control Center
TCP/IP	Telecommunication Control Program/Input
TEA/TEB	Triethyl-Aluminum/Triethyl-Borane
TMS	Thrust Measurement System
TOB	Test Operations Building
TTSC	Test & Technical Support Contractor

—U—

UHP	Ultrahigh Pressure
UPS	Uninterruptible Power Supply

—V—

VAC	Volts Alternating Current
VDC	Volts Direct Current
VJ	Vacuum Jacketed
VP	Valve P

ABBREVIATIONS

°C	degree Celsius	L	liter
°F	degree Fahrenheit	lb	pound
μ	micron (micrometer)	lbf	pounds force
μA	microampere	lbm	pounds mass
μin	microinch	lb/hr	pounds per hour
μs	microsecond	LH_2	liquid hydrogen
μV/D	microvolt/division	LHe	liquid helium
A	ampere	LN_2	liquid nitrogen
ac	alternating current	LOX	liquid oxygen
CO	carbon monoxide	M	million
CO_2	carbon dioxide	M-lb	million pound
dB	decibel	m/s	meters per second
dc	direct current	Mbyte	megabyte
ft	feet	Mbyte/s	megabyte per second
ft^2	square feet	mi	mile
ft^3	cubic feet	min	minute
ft^3/hr	cubic feet per hour	mm	millimeter
ft-lb	foot-pound	Mohm	megohm
ft^3/min	cubic feet per min	ms	millisecond
gal	gallon	MVA	megavoltampere
gal/min	gallons per minute	N_2	nitrogen
Gbyte	gigabyte	ns	nanosecond
GH_2	gaseous hydrogen	ppb	parts per billion
GHe	gaseous helium	ppm	parts per million
GHz	gigahertz	psi	pounds per square inch
GN_2	gaseous nitrogen	psig	pounds per square inch gauge
Gohm	gigohm		
GOX	gaseous oxygen	RP-1	rocket propellant 1
H_2	hydrogen	rpm	revolutions per minute
H_2O_2	Hydrogen Peroxide	s, sec	second
He	helium	scf	standard cubic feet
hp	horsepower	scfm	standard cubic feet per minute
Hg	mercury		
hr	hour	sps	samples per second
Hz	hertz	V	volt
in.	inch	Vac	volts alternating current
kHz	kilohertz		
ksps	kilo samples per second	Vdc	volts direct current
kV	kilovolt	wk	week
kVA	kilovoltampere		
kW	kilowatt		

REPORT DOCUMENTATION PAGE

Form Approved
OMB No. 0704-0188

1. REPORT DATE (DD-MM-YYYY)	2. REPORT TYPE	3. DATES COVERED (From - To)
16-04-2002		

4. TITLE AND SUBTITLE	5a. CONTRACT NUMBER
Test-Facility Capability Handbook, Fourth Edition, Revision 1 for off-site distribution	5b. GRANT NUMBER
	5c. PROGRAM ELEMENT NUMBER

6. AUTHOR(S)	5d. PROJECT NUMBER
Robert Bruce	
Paula Taliancich	5e. TASK NUMBER
	5f. WORK UNIT NUMBER

7. PERFORMING ORGANIZATION NAME(S) AND ADDRESS(ES)	8. PERFORMING ORGANIZATION REPORT NUMBER
Propulsion Test Directorate	NP-2001-11-00021-SSC

9. SPONSORING/MONITORING AGENCY NAME(S) AND ADDRESS(ES)	10. SPONSORING/MONITOR'S ACRONYM(S)
	11. SPONSORING/MONITORING REPORT NUMBER

12. DISTRIBUTION/AVAILABILITY STATEMENT
Publicly Available STI per form 1676

13. SUPPLEMENTARY NOTES

14. ABSTRACT

15. SUBJECT TERMS

16. SECURITY CLASSIFICATION OF:			17. LIMITATION OF ABSTRACT	18. NUMBER OF PAGES	19b. NAME OF RESPONSIBLE PERSON
a. REPORT	b. ABSTRACT	c. THIS PAGE			Robert Bruce
					19b. TELEPHONE NUMBER (Include area code)
U	U	UU		144	(228) 688-1646

Standard Form 298 (Rev. 8-98)
Prescribed by ANSI Std. Z39-18